What Every Engineer Should Know About Digital Accessibility

Accessibility is a core quality of digital products to be deliberately addressed throughout the development lifecycle. *What Every Engineer Should Know About Digital Accessibility* will prepare readers to integrate digital accessibility into their engineering practices. Readers will learn how to accurately frame accessibility as an engineering challenge so they are able to address the correct problems in the correct way.

Illustrated with diverse perspectives from accessibility practitioners and advocates, this book describes how people with disabilities use technology, the nature of accessibility barriers in the digital world, and the role of engineers in breaking down those barriers. Accessibility competence for current, emerging, and future technologies is addressed through a combination of guiding principles, core attributes and requirements, and accessibility-informed engineering practices.

FEATURES

- Discusses how technology can support inclusion for people with disabilities and how rigorous engineering processes help create quality user experiences without introducing accessibility barriers
- Explains foundational principles and guidelines that build core competency in digital accessibility as they are applied across diverse and emerging technology platforms
- Highlights practical insights into how engineering teams can effectively address accessibility throughout the technology development lifecycle
- Uses international standards to define and measure accessibility quality

Written to be accessible to non-experts in the subject area, *What Every Engineer Should Know About Digital Accessibility* is aimed at students, professionals, and researchers in the field of software engineering.

What Every Engineer Should Know

Series Editor: Phillip A. Laplante, Pennsylvania State University

What Every Engineer Should Know about Career Management
Mike Ficco

What Every Engineer Should Know about Starting a High-Tech Business Venture
Eric Koester

What Every Engineer Should Know about MATLAB® and Simulink®
Adrian B. Biran

Green Entrepreneur Handbook: The Guide to Building and Growing a Green and Clean Business
Eric Koester

What Every Engineer Should Know about Cyber Security and Digital Forensics
Joanna F. DeFranco

What Every Engineer Should Know about Modeling and Simulation
Raymond J. Madachy and Daniel Houston

What Every Engineer Should Know about Excel, Second Edition
J.P. Holman and Blake K. Holman

Technical Writing: A Practical Guide for Engineers, Scientists, and Nontechnical Professionals, Second Edition
Phillip A. Laplante

What Every Engineer Should Know About the Internet of Things
Joanna F. DeFranco and Mohamad Kassab

What Every Engineer Should Know about Software Engineering
Phillip A. Laplante and Mohamad Kassab

What Every Engineer Should Know About Cyber Security and Digital Forensics
Joanna F. DeFranco and Bob Maley

Ethical Engineering: A Practical Guide with Case Studies
Eugene Schlossberger

What Every Engineer Should Know About Data-Driven Analytics
Satish Mahadevan Srinivasan and Phillip A. Laplante

What Every Engineer Should Know About Reliability and Risk Analysis
Mohammad Modarres and Katrina Groth

What Every Engineer Should Know About Smart Cities
Valdemar Vicente Graciano and Mohamad Kassab

What Every Engineer Should Know About Digital Accessibility
Sarah Horton and David Sloan

For more information about this series, please visit: www.routledge.com/What-Every-Engineer-Should-Know/book-series/CRCWEESK

What Every Engineer Should Know About Digital Accessibility

Sarah Horton and David Sloan

CRC Press
Taylor & Francis Group
Boca Raton London New York

CRC Press is an imprint of the
Taylor & Francis Group, an **informa** business

Designed cover image: iStock Photo, Credit: mofles

First edition published 2024
by CRC Press
2385 NW Executive Center Drive, Suite 320, Boca Raton FL 33431

and by CRC Press
4 Park Square, Milton Park, Abingdon, Oxon, OX14 4RN

CRC Press is an imprint of Taylor & Francis Group, LLC

© 2024 Sarah Horton and David Sloan

ISBN: 978-1-032-26385-4 (hbk)
ISBN: 978-1-032-26386-1 (pbk)
ISBN: 978-1-003-28806-0 (ebk)

DOI: 10.1201/9781003288060

Typeset in Minion
by codeMantra

Epigraph

Accessibility is:

A core value, not an item on a checklist.

A shared concern, not a delegated task.

A creative challenge, not a challenge to creativity.

An intrinsic quality, not a bolted-on fix.

About people, not technology.

—Adapted from the Manifesto for Accessible User Experience*

---

* The Manifesto for Accessible User Experience was drafted in 2014 by UXPA London workshop attendees Chris Bailey, Graham Cook, Amber DeRosa, Dana Douglas, Yolanda González, Jack Holmes, Sarah Horton, Caroline Jarrett, David Sloan, Jennifer Sutton, Henny Swan, and Léonie Watson, and further refined with conference participants at WebVisions Chicago, A11yCamp Toronto, UX Lausanne, and UX Scotland in 2014–2015. The manifesto defines accessibility in terms of empowerment and opportunity, encouraging a creative, human-centered approach, where accessibility and disability inclusions are fundamental and essential to professional practice. We invite you to read and adopt the manifesto: accessibleux.wordpress.com/manifesto.

CONTENTS

SERIES STATEMENT

What every engineer should know includes an overwhelming catalogue of information. Regardless of discipline, engineering intersects most scientific fields and modern technologies. Practicing engineers, however, must also navigate managerial, socio-economic, and even political concerns. The engineer discovers soon after graduation that any curriculum omits important and thorny issues of daily practice—for example, problems concerning new technologies, scientific advances, business practices, legal implications, and team dynamics.

With the *What Every Engineer Should Know* series of concise, easy-to-understand volumes, every engineer can access primers on important subjects across a broad range of knowledge areas, including intellectual property, contracts, software, business communication, management science, and risk analysis, as well as more specific topics such as embedded systems design. These books are very accessible to every engineer, scientist, and technology professional and are necessary to remain competitive in this dynamic, global economy.

PREFACE

As an engineer involved in building digital products, we assume you want the products you're making to be useful and usable to people in your target audience. It's hard to think of a reason why anyone would invest time and energy in deliberately building a digital product to be difficult or impossible for people to use.

Unfortunately, it's all too common to encounter a digital product that's difficult or impossible for some people to use because it wasn't designed with accessibility in mind. A significant explanation for inaccessible digital products is a lack of accessibility awareness across product teams.

Accessibility may not be considered a core product quality to be integrated into decision-making and development. Perhaps there's an assumption that accessibility comes automatically or that it isn't necessary for the product's target audience. Maybe it's considered too complex or expensive to deliver.

This lack of awareness exists despite the emergence of digital accessibility as a key topic of discussion, a technical requirement, and an imperative for social inclusion and disability rights. The World Health Organization estimates that over 1.3 billion people, or 16% of the world's population, experience significant disability.[1] A growing number of people benefit from and require digital accessibility.

And yet, while awareness has increased, there's still plenty of evidence to suggest that accessibility levels of digital products are not where they need to be. A 2023 survey of 1 million website home pages by WebAIM found that each page had an average of 50 accessibility errors.[2] And although the number of pages with accessibility barriers reduced slightly from prior years, over 96% of the pages analyzed contained accessibility barriers. We have work to do.

But where to begin? As a topic at the intersection of technical quality and social inclusion, digital accessibility can easily become overwhelming, especially for people who are new to the subject. Do you need to know the details of your country's disability rights legislation? Do you need to become an expert user of every assistive technology that people who use your product might use? Do you need to be able to recite every constituent part of the Web Content Accessibility Guidelines?

Accessibility efforts can be easily derailed by uncertainty of where to begin to build knowledge and skills. It can be intimidating to suspect (or realize) that the product you're building has significant accessibility barriers. One response might be to focus on actions that may have little positive impact on disabled product users. Another response is paralysis and inaction. That's where this book comes in.

As a new volume and topic area for the *What Every Engineer Should Know* series, our goal for this book is to distill and present the core underlying principles and practices that engineers should know about digital accessibility. Regardless of the role you might perform, we want to help you grow accessibility knowledge and skills in a deliberate and progressive way, integrate accessibility into your engineering practices, and support others to do the same. To help tackle the subject, we've divided the book into two main sections:

- Part 1: Foundations of Digital Accessibility, covering the foundational concepts around digital accessibility—disability, inclusion, assistive technology—and providing an overarching framework for accessibility and your role as an engineer in delivering accessibility.
- Part 2: Methods for Engineering Digital Accessibility, focusing on the practice of engineering accessibility, from requirements through design, implementation, testing, documentation, and support.

Our aim is to provide you with a grounding in digital accessibility as a concept and practice. We explore ways that disability can influence how people use technology and how technology can respond to user accessibility needs. We cover how and when accessibility can be addressed in each stage of the product development lifecycle to optimize the chances of delivering a product that disabled people can successfully use.

And as authors, we don't speak to this topic alone. We've collectively spent over 40 years in the field, collaborating with colleagues, advocates, and users to advance digital accessibility and disability inclusion. Neither of us has lived experience of disability. While we have plenty to say about what every engineer should know, our views are limited by our domain of expertise and lived experiences. To provide more inclusive coverage of the topic, we asked several members of accessibility and disability communities to share their perspectives on what every engineer should know about digital accessibility. We provide their insights throughout this book as sidebars and are immensely grateful for their contributions.

There are topics that we don't cover. We don't cover the broader process of inclusive design for excluded and under-represented groups. Our focus is on disability, but we do note where digital accessibility can benefit inclusion of other groups.

We don't exclusively focus on web accessibility. While we present more examples from web accessibility than other platforms, the principles and approaches outlined are intended to be applicable to projects on other platforms.

And we don't provide much in the way of code. Our focus is on principles and best practices, with a few basic code examples as illustrations. You can apply the principles and best practices to your development environment and the coding language of your choice.

We hope this book serves as a helpful primer for engineers working across a range of platforms, using a range of development methodologies and tools, building digital products for a range of purposes, and working with and for a range of product teams and organizations. There are many excellent resources that dive more deeply into the social and technical aspects of accessibility and connect it to other dimensions of digital product quality, such as usability and user experience, security, responsiveness, and performance. We reference some of these resources as notes at the end of each chapter.

This book spans a range of topics related to computer science, including software engineering, human–computer interaction, web and mobile application development, and product and project management. This is intentional. Accessibility considerations apply regardless of the role you might perform in a digital product development. Everyone has an opportunity to make informed decisions that help ensure that accessibility is a priority.

We hope that upon finishing this book, you will have enough accessibility knowledge to be confident in progressing on your journey toward becoming an accessibility-aware professional, with the hunger to build on and apply that knowledge in your work moving forward.

Notes

1 World Health Organization – Disability Fact Sheet www.who.int/news-room/fact-sheets/detail/disability-and-health
2 The WebAIM Million webaim.org/projects/million/

ACKNOWLEDGMENTS

We are extremely grateful to everyone who contributed community perspectives and provided valuable feedback on drafts—Matthew Tylee Atkinson, Jonathan Avila, Yasmine Elgaly, Kate Kalcevich, Emily Ladau, Erich Manser, Jonee Meiser, Lē Silveus, and Makoto Ueki. We benefit from your knowledge and perspectives, and we know our readers will, too. We also thank David Swallow and Mark Sadecki for helpful guidance, encouragement, and feedback along the way.

We thank series editor Philip Laplante for recognizing digital accessibility as something every engineer should know, and Allison Shatkin, Chelsea Reeves, and the production team at Taylor & Francis and CRC Press for seeing the book through to fruition.

The contents of this book and the experiences that informed them are the result of many years of engagement with talented and generous colleagues. The people who have taught us, influenced us, and inspired us are too many to name individually—to anyone who's ever talked to us or written about accessibility, that means you, and we extend our gratitude. But we would like to pay particular tribute to Ashley Bischoff, Judy Brewer, Rebecca Bridges, Andy Coverdale, Steve Faulkner, Peter Gregor, Billy Gregory, Karl Groves, Hans Hillen, Ryan Jones, Patrick Lauke, Erin Lauridsen, Jonathan Lazar, Sarah Lewthwaite, Gez Lemon, Alan Newell, Adrian Roselli, Henny Swan, Annalu Waller, and Léonie Watson for their profound impact on how we approach digital accessibility. And a very special thanks to Mike Paciello, who made it all happen.

We are especially grateful to the people with disabilities who have participated in research studies over the years for giving their time and sharing their experiences and perspectives on how to make the digital world more inclusive.

Finally, we express our deepest gratitude for the ongoing love and support of our families, especially to Nico, Craig, and Finlay.

AUTHORS

Sarah Horton (she/her) has over 20 years of experience helping organizations create "born accessible" technology. She is an author of books, articles, and papers on designing technology to improve quality of life. She is current UX Strategy and Accessibility Lead at Harvard University, a Visiting Research Fellow at the University of Southampton, and an Invited Expert with Teach Access and the W3C's Accessibility Guidelines Working Group.

David Sloan (he/him) is Chief Accessibility Officer and UX Practice Manager at TPGi, a specialist digital accessibility services provider, and works with a range of clients to help them create accessible digital user experiences and build accessibility capacity in a sustainable way. He became interested in digital accessibility at the end of the 1990s as a postgraduate researcher at the University of Dundee, focusing on improving technology design for disabled and older people, and earned a PhD on web accessibility in 2006. While at Dundee, he taught classes on human–computer interaction and web design, co-founded the Digital Media Access Group, one of the world's first digital accessibility consultancy groups, and drafted the University's first accessibility policy.

CONTRIBUTORS

Matthew Tylee Atkinson (he/him) is a Principal Accessibility Engineer at TPGi and co-chair of the W3C's Accessible Platform Architectures Working Group, which helps other groups ensure that their specifications promote accessibility and carries out research and development of its own. Initially prompted by his vision impairment, Matthew has worked on academic accessibility projects, as well as with a multitude of clients in industry, and often presents at conferences and meet-ups on how people can make their products more inclusive. He also maintains some open-source accessibility projects, such as the Landmarks browser extension, which improves keyboard navigation for web pages.

Jonathan Avila, CPWA, (he/him) has over two decades of experience in the digital accessibility field, guiding organizations to create inclusive experiences that are usable to a wide range of people with disabilities. He is a member of the W3C's Accessibility Guidelines Working Group and the International Association of Accessibility Professionals. He has supported organizations across many different environments, including web, mobile, documentation, extended reality, kiosks, and gaming, to empower people with disabilities to live their best lives. He is currently the Chief Accessibility Officer at Level Access. At Level, he focuses time on testing methodology, thought leadership, and accessibility program policy to build an inclusive workplace.

Yasmine Elglaly (she/her) is Assistant Professor of Computer Science at Western Washington University (WWU). She founded the KIND (Komputing for INclusion, and Disability) research lab at WWU, dedicated to innovatively raising awareness about accessibility and disability through design and education. Yasmine explores diverse methods for imparting accessibility and disability knowledge within fundamental computer science and machine learning courses. Yasmine serves as the secretary of Teach Access executive committee and the co-chair of SIGCSE's universal design committee. She is also a Community Ambassador at WWU for supporting diversity, equity, and inclusivity.

Kate Kalcevich (she/her) started in 2001 as a digital accessibility practitioner. Since then, she's led design teams, managed products, and built accessibility programs. Kate is currently the Head of Accessibility Innovation at Fable, a leading accessibility company enabling the development of inclusive digital products. In her role at Fable, she focuses on helping large organizations practice innovative approaches to accessibility. Kate has lived experience with disability as a life-long hearing aid wearer.

Emily Ladau (she/her) is a passionate disability rights activist, writer, storyteller, and digital communications consultant whose career began at the age of 10, when she appeared on several episodes of *Sesame Street* to educate children about her life with a physical disability. Her writing has been published in outlets including *The New York Times*, *CNN*, *Vice*, and *HuffPost*, and her first book, *Demystifying Disability*, was published by Ten Speed Press, an imprint of Penguin Random House, in September 2021. Emily has spoken before numerous audiences, from the U.S. Department of Education to the United Nations. Central to all of Emily's work is harnessing the power of storytelling to engage people in learning about disability.

Erich Manser (he/him) began working in digital accessibility and disability rights advocacy in 2003, when he started noticing how his own vision loss was affecting his ability to use technology. He has worked to improve accessibility at leading tech companies like HP and IBM, and is now a Digital Accessibility Consultant at Harvard University. Erich is also an avid runner and triathlete, and he works to also help ensure equal, inclusive access to sports.

Jonee Meiser (she/her) discovered the barriers of digital inaccessibility when a high school student who frequented her tutoring center in 2012 struggled to use Kurzweil with their teachers' scanned handouts. Since then, she has developed a passion for making digital and educational spaces accessible to people with disabilities as an accessibility professional and instructional designer. She specializes in developing strategies to improve access to digital content, services, and products while emphasizing product management participation in digital accessibility. Currently, Jonee works as a Senior Accessibility Support Specialist at a digital educational publishing company in the Higher Education division. In her spare time, Jonee enjoys shuffling her two young children around to their activities and events with her husband.

Lē Silveus (they/them) is the CEO and Founder of Larunda Inc., an innovative venture championing inclusive digital experiences and neurodiversity at work. With a background in accessibility engineering, neurodiversity advocacy, and public leadership, Lē brings a diverse skillset and a passion for empowering marginalized communities. They are also a dynamic public speaker, leveraging their early career skills in Public Relations and Sales. Additionally, Lē is the visionary force behind Supported Living, a Texas-based 501(c)(3) nonprofit organization providing supported residential programs for neurominority and chronically ill adults. With their dedication to accessibility and community empowerment, Lē is driving positive change in the digital landscape and beyond.

Makoto Ueki (he/him) is a Web Accessibility Consultant in Japan. Makoto has been contributing to the JIS (Japanese Industrial Standard) Working Group and the W3C's Accessibility Guidelines Working Group since 2004. He is currently working on the WCAG 3 as an invited expert, and was chairman of the Web Accessibility Infrastructure Committee (WAIC) in Japan from 2012 to 2019. He is the first Japanese to be certified as an IAAP Certified Professional in Web Accessibility (CPWA). Makoto also co-organizes an online event to celebrate the Global Accessibility Awareness Day every year and the monthly A11y Tokyo meetup.

Part 1

Foundations of Accessibility

Technology is a source of both opportunities and challenges for people with disabilities. On the one hand, technology can mitigate the disabling effects of impairments. On the other hand, technology can produce barriers that limit or prevent disabled people from having full and equal access to programs, services, and activities. A solid accessibility practice allows engineers to capitalize on the opportunities and address the challenges. Part 1, *Foundations of Accessibility*, introduces the building blocks of digital accessibility, building awareness and understanding of who benefits from accessibility and in what ways, and overarching approaches to addressing accessibility needs.

- Chapter 1, *Introduction to Digital Accessibility*, defines digital accessibility and describes its beneficiaries and characteristics.
- Chapter 2, *Disability and Digital Inclusion*, focuses on establishing the core concepts of disability, disability discrimination, and the promise of digital inclusion.
- Chapter 3, *User Accessibility Needs*, covers a range of accessibility needs arising from disability, aging, and other impairments, as well as the situational and temporal nature of accessibility needs.
- Chapter 4, *Assistive Technology*, presents assistive technologies and accessibility strategies and how they manifest in both open and closed systems.
- Chapter 5, *Core Attributes*, focuses on characteristics such as flexibility and compatibility that make digital technologies particularly effective at meeting accessibility needs.

DOI: 10.1201/9781003288060-1

- Chapter 6, *Guiding Principles*, covers the principles, guidelines, and standards that underpin accessible design and engineering methods and practices.
- Chapter 7, *Accessibility in Practice*, brings it all together with guidance on how to integrate the foundations into professional practice.

Once you have read these chapters, you will have the foundational awareness and knowledge necessary to move on to the second part of this book, on methods for engineering digital accessibility.

1

INTRODUCTION TO DIGITAL ACCESSIBILITY

Objectives

In this chapter, we introduce accessibility and outline what it means in the digital context. We explore how accessibility features benefit a range of use cases and examine the critical role of design and engineering in advancing disability inclusion.

Once you're through this chapter, you should:

- Have a working definition of the concept and practice of accessibility.
- Understand how accessibility applies to digital products and services.
- Appreciate the broad range of beneficiaries of accessible digital products.
- Be familiar with the concept of the disability divide and your role in helping to close the divide through digital accessibility.

Introduction

As an engineer learning about creating accessible digital products, you are part of a movement that has evolved to advocate and fight for a fair society for people with disabilities, addressing years and years of exclusion. With an accessibility-informed practice, you have the opportunity to design and build innovative products that help create an accessible and inclusive digital world.

1.1 About Digital Accessibility

What do we mean when we talk about an accessible digital product? Does it mean conformance with accessibility standards or regulations? Does it

DOI: 10.1201/9781003288060-2

mean that the product has been tested using a screen reader? Does it mean that people with disabilities can use it? Does it mean having a plan and process in place to handle accessibility barriers when they're discovered? We expect most readers will have opened this book with some working definition of what is meant by the term accessibility, but we want to proceed with a clear, shared definition of what it is we're talking about.

1.1.1 Accessible

In some contexts, the term "accessible" may be used to describe whether systems are available to audiences more generally. "Is the system accessible?" could mean, is it turned on and available for use? Is it affordable? Is it physically reachable (or in, say, a locked room)? Is it available to people who have limited technology access? That's not the definition of accessibility we're addressing in this book.

The United States Department of Education's Office of Civil Rights (OCR), responsible for ensuring equal access to education in the US, provides a clear, precise, and helpful definition of "accessible" in the context of websites. The following definition is excerpted from the resolution agreement with the South Carolina Technical College System (SCTCS) "regarding the accessibility of its website to persons with disabilities, especially those requiring the use of assistive technology." In the agreement, the OCR defines "accessible" as providing disabled people with as full, equal, and independent access to SCTCS websites as nondisabled people.

> Accessible means a person with a disability is afforded the opportunity to acquire the same information, engage in the same interactions, and enjoy the same services as a person without a disability in an equally effective and equally integrated manner, with substantially equivalent ease of use. A person with a disability must be able to obtain the information as fully, equally, and independently as a person without a disability.[1]

In this way, we can think of "accessible" as we do other qualities, like secure, stable, and performant—a quality attribute of the product we're engineering that relates to its usability by people with disabilities.

We also want to be clear that as a quality, accessible is not a binary state, at least not without a significant level of context to qualify what we mean. "The system is accessible" is a meaningless phrase without some additional detail. Accessible to whom? Accessible for what purpose? Accessible when? Accessible where? In our experience, we often encounter the phrase in its negative form, for example, "This website is completely inaccessible!" There, we can at least understand that it's likely someone, somewhere, in attempting to perform a task, encountered a barrier that affected task completion. Born out of frustration, the claim extends beyond that specific experience. "Completely inaccessible" implies that no one can use the product for any purpose.

Overstating a negative state of a product might be counterproductive. If a product team hears their system described as "completely inaccessible," what will motivate them to fix it? If the team hears their system described as "having accessibility defects that stop blind customers from being able to buy a product," that helps clarify and specify the problem, and focus attention on where to make improvements.

1.1.2 Accessibility

With "accessible" defined as a quality of a product, we can think of the term "accessibility" as an activity or process, as well as a quality of a digital product. Accessibility as a process is deliberate attention to people with disabilities in any engineering activity that focuses on or includes users of the product we're engineering.

So accessibility as a quality of a system needs to be clarified in context. Who's affected? In what way? Under what circumstances? Accurately reporting the state of accessibility is an important aspect of engineering accessibility that we'll return to later in this book. For now, let's consider who benefits from accessible digital products—the people for whom we engage in accessibility activities.

1.2 Beneficiaries of Digital Accessibility

People with disabilities are the primary focus of engineering accessible products. There is also a very large additional audience who may not be considered disabled but who also benefit from accessible digital products because of accessibility needs relating to temporary impairment or their context of use. We'll explore profiles of disability, temporary, and situational impairments in the next chapter. Here, we emphasize the importance of making intentional efforts to minimize accessibility barriers in digital products and maximize ease of use. Digital products with accessibility defects can present barriers for people with disabilities, with no way around those barriers other than seeking help from other people. By contrast, someone who has a temporary or situational impairment may have other paths toward resolution that do not sacrifice their independence.

Although the number of people with situational and temporary impairments who benefit from accessibility features could be substantial, take care to avoid equating their needs with those of people with disabilities. Situational and temporary impairments will likely disappear at some point; disability most likely will not. Disability is a lived experience; situational and temporary impairments are for the most part transient inconveniences. This is why we emphasize the phrase "secondary beneficiaries" when we talk about nondisabled groups who benefit from accessibility.

The opportunity to build a product that helps reduce exclusion for people with disabilities, a historically marginalized group, should be a compelling motivation for any engineer who believes in the value of their work. Add to that the opportunity accessibility presents to make a product easier to use by more people in more situations. Considering the size of this audience and the fact that accessibility done right does not degrade the user experience for any group, we can assert that accessibility benefits everyone.

1.3 Context for Digital Accessibility

The focus of this book is accessibility as applied to people with disabilities, and about "digital" accessibility specifically. It's about ensuring disabled people can access systems, software, websites, apps, and other digital products. So what do we mean when we say digital products? In some contexts, they're referred to as ICTs, short for Information and Communication Technologies. We tend to think about them as anything that people interact with through a digital interface, which includes desktop software, mobile apps, websites, web applications, digital documents, kiosks, public access terminals, games, interfaces for smart devices... the list goes on.

Accessibility of digital products, digital accessibility, accessible products, accessibility—we'll use variations on these constructs throughout the book. But our focus is always on ensuring disabled people have full and equal access in the digital world and aren't blocked by accessibility barriers.

1.4 Accessibility Features

Accessibility features are familiar aspects of our everyday environment, including buildings, parks, pathways, programs, and media. For example, doors may open automatically when a sensor detects someone approaching. These doors make entrances accessible for people who have difficulty operating manual doors. You may have a physical disability and need automatic doors to get in and out of buildings. You may find them more convenient than manual doors, especially when your arms are full. You might not think twice about automatic doors.

But the professionals who designed and built automatic doors, elevators, ramps, signage, and other features to make an environment more accessible thought carefully about those features. They accounted for them in their designs, specified them in their plans and blueprints, and installed them according to standards and specifications. The businesses that run in those facilities continue to think about them, making sure they're in good working order and compliant with safety and accessibility regulations.

As a software engineer, you are responsible for the digital equivalent of doors, passageways, transport, and signage. Like the designers, architects, and builders of the physical environment, it's up to you to make the digital world safe and accessible for everyone, including disabled people. For example, a web-based application is the equivalent of a digital entryway, with sign-up and login features that grant users access to goods and services. Links are the transport, allowing users to navigate from one feature to another. And headings and labels are the signage, telling users where they are and directing where they can go. Like the everyday accessibility features of the built environment, when digital accessibility features are included in mainstream technology, users might depend on them, find them convenient, or not be aware of them at all.

For accessibility features to be present in the digital world, professionals who design and build digital products must pay deliberate attention to accessibility, accounting for accessibility features in specification, design, implementation, and testing, and maintaining compliance with accessibility regulations. When digital product teams don't explicitly include accessibility as a requirement, chances increase that accessibility features will be overlooked and forgotten about. Accessibility gains will be lost at the expense of other product objectives, such as meeting a particular look and feel, or using a particular design pattern for a user interface component. Accessibility may be added to a product after-the-fact and at a disadvantage. When accessibility features are addressed as remediation, the result is usually suboptimal for everyone.

Fortunately, the digital environment has inherent features that make it conducive to accessibility. For example, let's consider adaptability. Hard-copy books are inflexible by nature; what you see is what you get. If you want a large print book, you need to buy a large print book. If you want an audio book, you need to buy an audio book. With digital books, there are seemingly limitless adaptations users can make *to the same book* in terms of how they consume the information, such as changing the text and background colors, text size, or having the text read aloud using text-to-speech software. Digital content and functionality are multi-dimensional, with content, structure, and presentation on different dimensions that can be experienced in different ways. And there are solid principles and guidelines we can follow that allow us to capitalize on the features of the digital environment in a way that maximizes its potential for accessibility and disability inclusion.

1.5 Closing the Disability Divide through Digital Accessibility

One reason digital accessibility is something every engineer should know is that, without deliberate care and attention, we risk exacerbating a societal divide. We risk creating products that disabled people cannot use, which then creates a divide, or rather, many divides, between people who have accessibility needs and all the facets of daily life that are mediated by digital products. We can think of this as the "disability divide."

You may be familiar conceptually with the digital divide, where access to digital products is not equally distributed across countries and within societies, creating a digital divide between those who have access and those who do not.[2] You may be personally affected by the digital divide. You may know people who are affected. These are critical considerations for engineers who are building digital products. And while many people with disabilities are among those who do not have access to digital products, and while in some cases, products that would address accessibility needs are out-of-reach financially and otherwise from people who would benefit from them, addressing the *digital* divide is not the focus of this book.

Addressing the *disability* divide, where disabled people are divided from nondisabled people in their access to essential technology-mediated activities like education, healthcare, employment, and government—this is the divide we seek to close with digital accessibility. To emphasize a critical concept—people with disabilities can and do successfully use technology products when they're designed appropriately. Problems arise when technology is not designed with disability in mind, not because it's impossible to provide accessible technology.

So how can engineers help to close the disability divide? We do that by striving toward digital inclusion, which we focus on in the next chapter, *Disability and Digital Inclusion*.

Every engineer should know… inaccessible software does not only eliminate or limit access, but it can also cause harm to some users.

By Yasmine Elglaly

Inaccessible software prevents users from accessing the digital information or the functionality it provides. This can result in limited or no access to educational services in the case of a learning management system or job search functionality in the case of an inaccessible job search platform. Consequently, the impact of inaccessible software on the quality of life of people with disabilities goes beyond the inability to access or operate the software; it is about the significance of that software in enabling them to be effective members of society. With accessible software, people with disabilities have an equitable opportunity to participate in society, including economical and political activities.

Inaccessible software can cause harm by excluding certain users, but the consequences of discrimination are even greater with AI-based systems. These systems have the potential to magnify harm and make disability bias more systematic. For example, AI-based systems may discriminate against people with disabilities by excluding them from advertisements for job positions and insurance services, and face, body, and speech recognition systems may not work effectively for them. It is essential for software engineers to actively prioritize inclusion in AI-based systems. Inclusive datasets are crucial and should ensure equitable representation of people with disabilities. Additionally, models must be trained to equally serve people with disabilities to ensure that these systems are fair and inclusive.

It may seem unlikely, but the design of a software's user interface or interaction can have a physical impact and cause harm. Certain features of software, especially those affecting users with neurological conditions like epilepsy, can pose health risks. Accessibility flaws can also negatively impact the well-being of neurodivergent users. Therefore, it's crucial to follow accessibility guidelines to protect the health of all users.

Although accessibility guidelines don't address neurodiversity extensively, the WCAG 2.2 guidelines include coverage for photosensitive and motion-sensitive users in Guideline 2.3 Seizures and Physical Reactions.[3] For instance, one guideline states that nothing on display should flash more than three times per second. This guideline, although marked as level AAA and not required by law to be implemented, is crucial for the well-being of neurodivergent users.

It's important to understand that a violation of an accessibility guideline isn't just a minor bug. An accessibility violation can render an app unusable and actively harm neurodivergent users. Therefore, it's essential to prioritize accessibility in software design to ensure the safety and well-being of all users.

Takeaways

As an engineer, you should:

- Adopt accessibility processes and practices in order to build accessible products.
- Prioritize user accessibility needs when building digital products; avoid causing harm.
- Appreciate the accessibility features inherent to the digital environment and use them to their best advantage in product design and development.
- Champion digital accessibility and disability inclusion.

Notes

1 United States Department of Education, Office of Civil Rights (2013) *Resolution Agreement, South Carolina Technical College System OCR Compliance Review No. 11-11-6002.* www.ed.gov/about/offices/list/ocr/docs/investigations/11116002-b.html

2 M. L. Pertegal-Felices, D. Marcos-Jorquera, A. Jimeno-Morenilla, R. Gilar-Corbi and H. Mora (2020) Training Future ICT Engineers in the Field of Accessibility and Usability: A Methodological Experience. *IEEE Access,* vol. 8, pp. 65812–65820.

3 W3C (2023) *Understanding Guideline 2.3: Seizures and Physical Reactions.* www.w3.org/WAI/WCAG22/Understanding/seizures-and-physical-reactions.html

2

DISABILITY AND DIGITAL INCLUSION

Objectives

In this chapter, we explore definitions of disability, and why an understanding of disability and disability rights is so important to digital accessibility. We explore what it means to design for people in the broadest sense, moving beyond notions of typical, average, or normal to a more accurate representation of human diversity and lived experience. We show how attention to accessibility and meeting diverse needs can result in innovative features and services.

Once you're through this chapter, you should:

- Be familiar with conceptual models of disability and disability rights.
- Understand how disability rights are affected and protected online.
- Appreciate how attention to accessibility can help protect disability rights and produce quality products and services.

Introduction

Digital products can help reduce or remove barriers experienced in everyday life by people with disabilities. To be an accessibility-aware engineer involves developing an understanding and appreciation of who benefits from accessibility and the implications of barriers that digital products can present. Disability is often misunderstood and mischaracterized, which has led to historical patterns of discrimination and sometimes misguided efforts to address disability inclusion. As a powerful illustration, Alice Wong's book, *Disability Visibility,* collects a diverse set of contributions from disabled people about their lived experiences, celebrating disability culture while also articulating the impact of discrimination.[1]

Understanding models of disability, the ways in which disability exclusion can occur, and the efforts undertaken to address discrimination provides a platform for productively collaborating with disabled people to build digital products and experiences that support efforts to create a more disability-inclusive world.

2.1 Models of Disability

Models of disability are helpful in understanding the nature of disability and its implications for you as an engineer. You may identify as a disabled person or a nondisabled person, or you may not have given much thought to what disability means. Models of disability are helpful for understanding different ways disability is defined and approached. They may help you contextualize disability and what it means to you as an engineer who is committed to disability inclusion. Here we explore three prominent models: the medical model, the social model, and the human rights model.

But before we describe the models in more depth, we, as authors, must first declare our biases. We are two people who identify as nondisabled and who have been dedicated to digital accessibility since the early days of the web. We believe that design and technology have enormous potential to reduce barriers to inclusion and close the disability divide. We have personally benefited from access to technology and the creativity that it affords—not least of which is writing this book! We are excited by the professional and personal challenge of using design and engineering to open opportunities to the broadest range of people as possible. We are committed to exploring ways of leveling the playing field so that everyone can participate fully in society, relationships, and existence. We do not come to that view from some altruistic ideology. We know the broad benefits of diverse perspectives and participation. As people whose decisions affect others in significant ways, we take our responsibility seriously and seek to make decisions that are the most inclusive. We don't always get it right, but those are our guiding motivations: to do no harm and to open up opportunities and access so that we all benefit.

2.1.1 Medical Model

As we explore the medical model of disability, you'll understand our need to declare our bias. With the medical model of disability, the locus of the problem of disability is the person. The medical model focuses on medical conditions and impairments as problems to be solved. In other words, the medical model problematizes the person and frames the person as someone in need of intervention, or "fixing."

In a modern society that is largely engineered, the medical model deflects responsibility for disabling conditions and situations away from engineers and therefore has limited place in this discussion of design and technology, which together are enablers. Imagine if access to public transportation was predicated on the ability to navigate stairways, and access barriers to using public transportation were written off as unsolvable problems due to the medical conditions of those who experienced them. This is an extreme example to make a point, even though we recognize that public transportation today still presents significant accessibility barriers. The medical model is still prevalent in some thinking around disability. It's important to mention it here, but only to illustrate its limitations. When accessibility decisions are guided by a medical-model view, they are likely to be narrowly focused and not inclusive. For example, efforts may focus on supplementing a product built for nondisabled people with separate accommodations for disabled people, perpetuating separation and exclusion from mainstream life.

We recognize that design and engineering play a huge part in whether or not people are disabled by the built and digital world. We also recognize that engineering produces technologies designed to minimize or mitigate the experience of disability. We cover some enabling technologies in the chapter on *Assistive Technologies*.

2.1.2 Social Model

On the other hand, under the social model of disability, responsibility for disability rests squarely on how society and our world are designed and constructed. With the social model, human difference is not innate but constructed. It's society and our built world that disables people. The social model is well represented in identity-first language, with "disabled people" rather than "people with disabilities," which we discuss more in depth below.

> This social model, an idea rooted in the disabled people's movement, entails a political repurposing of the idea of "disability," whereby that term is used to describe the socially created disadvantage and marginalization experienced by people who have (or are perceived to have) "impairments." At its heart is the distinction between socially created exclusion and disadvantage on the one hand and the particular mind and body traits of individuals on the other.[2]

Although awareness of the social model has grown in recent years, you might find that it's relatively absent from organizational decision-making. An accessibility practice supports the social model of disability, where everyone has a role to play in ensuring equal access and participation for people who have accessibility needs, including disabled people, older people, and people with temporary or situational impairments.

2.1.3 Human Rights Model

Similarly, the human rights model of disability locates the "problem" of disability in society rather than the person. Under the human rights model, the limiting aspect of disability arises from society's tendency to view people with disabilities as objects rather than as people with inherent rights and self-worth, and therefore to assign negative valuations about significance and worth.[3] Considering accessibility and disability concepts under the human rights model, we are all equally important, relevant, and useful as learners, teachers, designers, developers, end-users, and businesses. We are all equal stakeholders and participants in designing, engineering, and using digital technologies. We are all responsible for digital accessibility.

The social and human rights models of disability challenge all of us to step out of our known world and understand how we might be disabling others through our actions, attitudes, and the things we produce. Each of us has ways of being and doing that don't work for others. As engineers, if we design for ourselves, we will inevitably disable others, regardless of our own disability status. The way forward is to work to break free from limiting concepts and continuously learn and improve our understanding of others, so we can design and build successful products that improve lives in the broadest sense possible.

2.2 Disability Language and Representation

A deeper understanding of disability involves an awareness of how disability is represented in everyday life. In particular, we want to look specifically at the language of disability and how disability is represented in media, which are both significant and dynamic areas of activism. Given digital accessibility's close connections to disability rights and social inclusion, it's important to understand how to ensure that disability is addressed and represented positively and sensitively, so that language and representation do not get in the way of progress.

The language of disability is a constantly evolving phenomenon, shaped by changing cultural norms and differing across geographical locations. What's acceptable in one country might not be appropriate in another country. It's possible for a product team to spend a lot of time in debate, exploring the nuances of appropriate language of disability. Language use is incredibly important, but you might find that some of that time could have been spent fixing accessibility issues and building accessible digital products—which is arguably more impactful in the long term. That said, there are certain guiding principles that you can follow to ensure that when discussing and representing disability and accessibility, for example, in discussions, training and marketing materials, and product documentation, you do so in a way that minimizes the chances of causing offense or undermining accessibility and disability inclusion efforts.

There is a long-running debate over the use of "people-first" language versus "identity-first" language—in other words, whether to reference "people with disabilities," with disability status an attribute of the individual, or "disabled people," which positions disability status as a primary aspect of identity. This conceptual difference is connected to the wider debate around the social and human rights models and medical models of disability, as discussed above. Perspectives and preferences on approach vary according to factors such as global location, disability group, and individual identity. Unlike our bias toward social and human rights disability models, we do not take a position in this book on language, but rather use both constructs throughout. Both approaches have compelling arguments and promoters, and ultimately the most important decider is individual and group preferences. The best approach is to do research, ask people for their preference, and seek to understand the approach that is most acceptable based on your context. And we encourage you always to refer to disability directly and avoid euphemisms like "differently abled" or "special needs."

There are a few things you should avoid when trying to use disability-inclusive language:

- **Terms that are recognized to be offensive.** If you are unsure or need to ask whether a disability-related term is offensive in your language, it's likely that it will be. When in doubt, use an alternative term.
- **Terms that "other" people with disabilities.** If you use language that implies that you're talking to nondisabled people, then you naturally exclude anyone who identifies as disabled. When addressing an audience, "Those of us who are deaf may not hear the alert" is inclusive language. "You and I hear the alert, but people who are deaf may not" is not. Pay close attention to your assumptions about who is in the room and who isn't.[4]
- **Language that references legal status.** Focusing on laws in disability language can have the effect of categorizing people with disabilities as a legal compliance problem rather than as people you are trying to support. For example, avoid using language like "ADA customers" (referring to people covered by the Americans with Disabilities Act in the United States). Instead, include disabled people as customers and users who share the needs and wants of other customers.

Every engineer should know… how to talk about disability — an excerpt from *Demystifying Disability*.

By Emily Ladau

The choices that we, as disabled people, make about how we describe and define ourselves are deeply personal, and each of us has our own preferences. The way people who have a disability talk about their disability is their choice—I cannot stress this enough. We all need to respect these choices, even if we're also disabled and someone else's choices are different from our own. With that in mind, let's talk about disability.

There's one thing we should address right away—we need to stop using the word handicapped. It's an outdated term that's fallen out of favor with most disabled people, and, quite frankly, it makes my skin crawl. Occasionally, I'll find myself saying things such as "handicapped bathroom" or "handicapped entrance" because old terminology dies hard. But here are better words to use. Talking about a person? "Disability" is better than "handicap." Pointing out a parking spot with the blue lines? It's "accessible" parking. Now that we've settled that, you'll notice throughout this book that I switch between the terms "person with a disability" and "disabled person." That's partly because I like to shake things up a bit in my writing, but it's mostly because these terms honor two of the main ways of referring to disability: person-first language (PFL) and identity-first language (IFL).

Person first language (PFL) does just what it says: it puts the word "person" first before any reference to disability is made. This type of language is all about acknowledging that human beings who have disabilities are, in fact, people first, and they're seen not just for their disability. So when using PFL, you might say "person with a disability" or "person who has Down syndrome" or "people who use wheelchairs." The logic here is that disability is something a person has, rather than who they are, so by separating any mention of disability from the person and putting it second, you're showing you respect the personhood of someone with a disability.

Identity-first language (IFL) is all about acknowledging disability as part of what makes a person who they are. So when you use IFL, you might say "disabled person" or "blind person" or "Autistic people." In this case, disability isn't just a description or diagnosis; it's an identity that connects people to a community, a culture, and a history.[5]

From the excellent book, *Demystifying disability: What to know, what to say, and how to be an ally*, by Emily Ladau. Included with permission of the author.

2.3 Disability Discrimination

Disability discrimination occurs when a disabled person is treated differently and less favorably due to their disability. Architectural barriers cause disability discrimination, for example, when a customer with a mobility impairment is unable to enter a building due to steps. Communication barriers cause disability discrimination when a person with speech loss can't access a service because it uses an automated phone system operated by voice recognition technology. Digital barriers cause disability discrimination when a person who uses a screen reader can't access website content and functionality due to accessibility defects. Because of the prevalence of disability discrimination, society must put measures in place to protect disabled people from exclusion and other harms, and laws exist in many countries around the world to protect the rights of people with disabilities not to encounter discrimination.

2.3.1 Laws, Policies, and Regulations

Rules and regulations can be helpful when the best path forward is not clear or intuitive. As a society, we create rules when there's a need—when the lack of rules is either causing harm or could potentially cause harm, and when those who are responsible are not sufficiently aware or motivated to act with integrity and care. With digital accessibility, the right path forward may be obscured by limiting attitudes and beliefs and a lack of experience and understanding. Looking at relevant rules and requirements can be a helpful source of clarity about what's needed.

Many locations throughout the world have laws, policies, and regulations that address digital accessibility and disability inclusion. This framework protects the rights of disabled people to participate fully and equally in the digital world. Here we cover a few key examples:

- The United Nations Convention on the Rights of Persons with Disabilities (UN CRPD) is a global human rights treaty reaffirming the rights of people with disabilities to "enjoy all human rights and fundamental freedoms." Article 9 on accessibility includes directives to take measures to ensure equal access to information and communication technologies (ICT), with instruction to "promote the design, development, production and distribution of accessible information and communications technologies and systems at an early stage, so that these technologies and systems become accessible at minimum cost."[6]
- In the United States, federal antidiscrimination laws address digital accessibility, including Section 508 of the Rehabilitation Act, the Americans with Disabilities Act (ADA), and the 21st Century Communications and Video Accessibility Act (CVAA), among others. Many state and local governments also have laws and policies aimed at promoting equal access to digital resources for people with disabilities.
- In Canada, the Accessible Canada Act (ACA) is country-wide legislation that provides a framework for developing, implementing, and enforcing accessibility standards, and the Accessible Canada Regulations specify how federally regulated entities must operate in

meeting their obligations. Provinces and territories also have laws, such as the Accessibility for Ontarians with Disabilities Act (AODA), which specifies standards.

- The European Union's Charter of Fundamental Rights provides a basis for disability rights legislation for EU member states. The EU has also passed directives that require member states to adopt legislation and regulations addressing the accessibility of digital products in the public and private sectors.
- The United Kingdom's Equality Act includes protections for people with disabilities alongside other protected groups.
- Australia's Disability Discrimination Act is one of the earliest laws protecting the rights of disabled citizens from discrimination based on having a disability.

Some disability rights laws extend protections to include people with chronic physical or mental illnesses, and nondisabled people who are perceived as having a disability—reflecting the insidious nature of discrimination that sadly still exists.

Legislation protecting the rights of people with disabilities is generally civil or human rights legislation and typically defines those rights in high-level terms, along with the scope of application of the law—for example, whether it applies to publicly funded organizations, private organizations, or both. With a few exceptions, disability rights legislation generally does not specify detailed requirements for building digital products—that is left to standards and regulations that may be associated with or referenced by a law or in supporting guidance. Disability advocate and lawyer Lainey Feingold maintains a comprehensive list of relevant legislation, regulations, and standards from around the world.[7]

We (the authors) are not lawyers, and this is not a book about law, nor does it offer legal advice. So we'll limit ourselves to note that disability rights laws are likely to place obligations on your organization as an employer of disabled people, a provider of products and services that disabled people may use, or perhaps as a recipient of public or government funds. They may also place obligations on customers of digital products that you build, for example, if you sell products to public sector organizations, universities, or government organizations. In turn, you may work for an organization that has policies related to accessibility and disability inclusion, such as accessibility and procurement policies or employment policies. Laws and policies help organizations meet legal obligations and manage risk, as well as contribute positively to an inclusive society. Those same objectives make their way into codes of professional practice, with codified rules that aim to govern the profession, maintain necessary levels of competency, and promote safe practices.

Professional requirements

Membership in your professional organization might require that you operate within a code of ethics that includes nondiscrimination. For example, both ACM and IEEE have professional codes that support accessibility as a requirement in digital product development.[8,9]

For example, the following excerpt is from principle 1.4, *Be fair and take action not to discriminate*, from the ACM Code of Ethics and Professional Conduct:

> Computing professionals should foster fair participation of all people, including those of underrepresented groups. Prejudicial discrimination on the basis of age, color, disability, ethnicity, family status, gender identity, labor union membership, military status, nationality, race, religion or belief, sex, sexual orientation, or any other inappropriate factor is an explicit violation of the Code. Harassment, including sexual harassment, bullying, and other abuses of power and authority, is a form of discrimination that, among other harms, limits fair access to the virtual and physical spaces where such harassment takes place.
>
> The use of information and technology may cause new, or enhance existing, inequities. Technologies and practices should be as inclusive and accessible as possible and computing professionals should take action to avoid creating systems or technologies that disenfranchise or oppress people. Failure to design for inclusiveness and accessibility may constitute unfair discrimination.

2.4 Accessibility Barriers

In the physical world, many older buildings and environments were not built with an awareness of accessibility. They were designed and built in a time when demand for accessibility was less mainstream and was poorly regulated or not regulated at all. Some buildings have been retrofitted, with varying degrees of success, to make them more accessible to disabled people. By contrast, most modern buildings are built according to accessibility specifications, like the ADA Standards for Accessible Design in the United States, with requirements for accessibility features such as elevators, ramps, tactile signage, and hearing induction loops.[10]

In the digital world, when we build products from scratch, we have a responsibility to make them accessible. When we don't do that, retrofitting digital products can, like in the physical world, be expensive and not reliably effective in removing barriers. What do accessibility barriers look like in the digital world? Barriers may take a range of forms, such as:

- A hotel booking mobile app that relies on color perception to distinguish between hotels with available rooms and hotels with no rooms available.

- A retail website that includes rapidly flashing content advertising special offers.
- A desktop software application for managing warehouse logistics with a menu system that can only be operated using a mouse.
- A kiosk providing a means to book and collect train tickets that has no speech output capability.
- A point-of-sale touchscreen with small, tightly spaced controls.

All of these digital products contain accessibility barriers that hinder or prevent some people from using them for their intended purpose. In each case, there are straightforward steps that could have been taken during design and development to avoid barriers being introduced.

2.4.1 Origins of Accessibility Barriers

Accessibility barriers in digital products are, in our experience, rarely introduced with the deliberate intention of excluding people. Instead, they are most likely to result from one or more of the following situations:

- Accessibility has not been considered a sufficiently high priority by the organization commissioning or building a digital product to ensure that it is given due attention.
- Planning phases of product design and development have not sufficiently taken accessibility needs into account.
- Accessibility has not been specified in product requirements, or if it has, in insufficient detail.
- The team building the product collectively has insufficient awareness, knowledge, and skills to enable them to make accessibility-informed decisions.
- There are insufficient numbers of people in the team to cover all accessibility-related tasks.
- The team has an accessibility specialist, but they have insufficient authority to ensure that their recommendations are followed.
- The tools used by the team to build and test the product do not support accessibility.
- Schedule constraints mean that accessibility-related activities are dropped.
- Accessibility testing does not start until very late in the project lifecycle.
- Accessibility bugs are not fixed due to a lack of time and resources.

Digital products are rarely shipped with accessibility barriers because it is technically impossible to avoid introducing those barriers. Far more common is that accessibility was insufficiently prioritized when building the product. Product and project management might not have understood accessibility as an essential feature, and therefore, did not treat accessibility defects as blockers. The development team may not have had sufficient knowledge and skills to design and implement accessibility. Given these process-based origins of accessibility barriers, we can approach accessibility as a process improvement effort, one that can be guided by inclusive design.

2.5 Digital Inclusion

As an engineer designing and building digital products, you have a responsibility to avoid introducing barriers that might inhibit or prevent disabled people from using digital products. We can think about digital accessibility as an intentional effort to meet our responsibility for avoiding creating products that exclude some users. The danger of a focus on legal compliance is that it can lead to bare-minimum efforts to satisfy the letter of a law, or tick off items on a checklist, without true engagement with solving a design challenge in the most effective way. Thankfully, we can also think about accessibility as an opportunity to engineer better digital products for disabled people and for a broader audience.

2.5.1 Inclusive Design

The principles and practices of designing for diversity are referred to as inclusive design and involve a focus on understanding diverse user needs and diverse interaction methods. Adopting a practice of inclusive design means improving quality and usability for all users, especially groups who have historically experienced exclusion and discrimination, whether that is due to race, gender, sexual orientation, economic status, disability, age, religious beliefs, or a range of other characteristics.

Since this book focuses on disability, it's helpful to realize that an intentional focus on addressing the needs of people with disabilities can provide inspiration and insights that lead to new and innovative products benefiting a broad audience. Accessibility knowledge and skills can provide a path toward interesting employment. An inclusive design practice focuses attention on the needs of people, helps improve product quality, and matures your professional practice—so the practice of inclusive design is something we'll return to regularly in this book.

2.5.2 Quality

Accessibility is a quality attribute, and inclusive design improves quality. If a supposed inclusive design process leads to a reduction in quality for some users, then that's not true inclusion. The way you design and engineer solutions to accommodate diversity can lead to products that are more resilient, more robust, and more flexible in how they are used. The characteristics of an accessible digital product align well with other objectives that may be seen as priorities for the product, for example:

- Many principles of accessible web and document design are also recognized as supporting search engine optimization (SEO), as accessible digital documents tend to have the semantic information and alternatives for non-text content that enable effective indexing.
- Accessibility best practices can also help address challenges relating to low-bandwidth access, catering as they do to situations where a user does not have access to images or multimedia and encouraging the use of native user interface elements over elaborately scripted custom elements. In turn, this helps create digital products that have more global usage prospects, including for people who do not have access to high-end devices and robust data connections.
- Mobile-friendly design has significant overlap with accessible design—the usability challenges that a smaller screen and touchscreen input can present can, to a significant extent, be addressed by the same design best practices that address visual, cognitive, and dexterity limitations.

2.5.3 Innovation

The nature of accessibility's focus on accommodating diversity in input and output channels, and in interaction methods and environments, brings opportunities to create digital products that can be used in new and often unexpected ways and that are more robust and resilient when used in constraining situations. A focus on accessibility brings process adjustments that can also help bring greater clarity to understanding a product's requirements and increase the chances of building a product that works for more people in more situations without expensive redesign or retrofitting.

You might feel that a commitment to accessibility places extreme constraints on design and engineering creativity and innovation. The reality is that the opposite can be true. In the words of accessibility advocate Léonie Watson, accessibility is "a creative challenge, not a challenge to creativity."[11] Thinking about accessibility can help lead to innovative solutions. There are many examples where a product designed to meet specific user accessibility needs has led to innovation that benefits other audiences, for example:

- The cassette recorder was developed to provide a way for blind people to access written content by enabling tape recording, storing, and playing content in audio format. Spoken audio was soon augmented by music. Cassette recording then became a means to record, store, and play video content and digital data.

- The logic behind predictive text technology was created in an effort to help people with speech and communication difficulties communicate more effectively and efficiently. The emergence of mobile phones as text-based communication devices, with their restricted keyboard input options, brought predictive text technology to a wider audience.
- Speech input technology developments were driven in part by the need to provide alternative ways for people with physical disabilities to interact with a computer. Today, speech input technology has evolved to the point that it's a reliable and common way to interact with devices ranging from smartphones to home assistants.

In each of these examples, the primary initial focus of the innovation was on solving an accessibility challenge. The subsequent adoption of the technology may not have been predicted by its inventors, but the benefits the innovation has brought to the world should not be underestimated. Tim Brown, founder of the design agency IDEO, argued in his book, *Change by Design*, that designers should focus on populations at the margins, people typically neglected by designers when considering user needs. "By concentrating solely on the bulge at the center of the bell curve, we are more likely to confirm what we already know than learn something new and surprising."[12] In the context of accessibility, a focus on understanding and solving for user accessibility needs can lead to innovative technologies that are essential to everyday life.

We want to be careful here to avoid implying that developing accessible digital products is only worthwhile when there is an audience beyond the disability group that directly benefits. This is demeaning and disrespectful to disabled people. The argument that "accessibility is not just about people with disabilities" is often used to justify a return on investment in accessibility efforts, as if designing something that people with disabilities can use is insufficient motivation on its own. Describing the positive impact of accessible digital products on other audiences can certainly be persuasive, and in some cases, it may be necessary to convince some people, but it's important not to appear to be trivializing or downplaying the benefits for people with disabilities when making an argument for greater focus on accessibility.

We see value in examining the benefits of an inclusive design approach because exploring accessibility in a broader context can provide additional fuel and momentum to accessibility and digital inclusion efforts. But our focus is on meeting the accessibility needs of people with disabilities, and we will explore these needs further in the next chapter.

Takeaways

As an engineer, you should:

- Embrace the social and human rights models of disability and take responsibility for minimizing the disabling effects of whatever you produce.
- Examine your attitudes and biases around disability and seek to expand your awareness and appreciation of a broad range of lived experiences.
- Pay close attention to how you talk about disability and how you reference and represent disability in your work and work products.
- Be aware of relevant laws for your location and operate in compliance with the accessibility laws and regulations that govern your context, including professional codes of conduct.
- Prioritize meeting accessibility standards when designing and implementing digital products and avoid creating accessibility barriers.
- Champion accessibility and inclusive design in your professional practice as an opportunity for excellence and innovation.

Notes

1 A. Wong (ed.) (2020) *Disability Visibility—First-Person Stories from the Twenty-First Century*. Vintage Books.
2 A. Lawson and A. E. Beckett (2021) The Social and Human Rights Models of Disability: Towards a Complementarity Thesis. *Int J Human Rights*, vol. 25, no. 2, pp. 348–379.
3 G. Quinn and T. Degener (2002) The Moral Authority for Change: Human Rights Values and the Worldwide Process of Disability Reform. In *Human Rights and Disability: The Current Use and Future Potential of Human Rights Instruments in the Context of Disability*, eds. G. Quinn and T. Degener (United Nations, 2002), 13, 14.
4 S. Horton and E. Lauridsen (2023) We Need to Talk about How We Talk About Accessibility. *J User Exp*, vol. 18, no. 3, pp. 105–112.
5 E. Ladau (2021) *Demystifying Disability: What to Know, What to Say, and How to be an Ally*. Emeryville: Ten Speed Press.
6 United Nations (2006) *Convention on the Rights of Persons with Disabilities: Article 9—Accessibility*. social.desa.un.org/issues/disability/crpd/article-9-accessibility
7 L. Feingold (2023) *Global Law and Policy*. www.lflegal.com/global-law-and-policy
8 Association for Computing Machinery (2018) *ACM Code of Ethics and Professional Conduct*. www.acm.org/code-of-ethics
9 IEEE (2014) *IEEE Code of Ethics*. www.ieee.org/about/corporate/governance/p7-8.html

10 United States (2010) *ADA Standards for Accessible Design.* www.ada.gov/
    law-and-regs/design-standards
11 *Manifesto for Accessible User Experience.* accessibleux.wordpress.com/
    manifesto
12 T. Brown (2009) *Change by Design: How Design Thinking Transforms
    Organizations and Inspires Innovation.* New York: Harper Collins.

3

USER ACCESSIBILITY NEEDS

Objectives

In this chapter, we introduce the groups of people who benefit from accessibility and provide a foundation in the user accessibility needs that must be met when building and providing digital products. We explore how disabilities and other impairments can create accessibility needs for people using digital products.

Once you're through this chapter, you should:

- Understand the range of accessibility needs of people with disabilities when using digital products.
- Appreciate how different groups of people can benefit from digital accessibility efforts.
- Understand how technology use can be impacted by different types of impairments.

Introduction

As an engineer, you might be tempted to go straight to learning about accessibility specifications and standards and to use this information to integrate accessibility into the digital products that you build. But it's important to take a step back to understand more about whom you're supporting and what their needs and goals are. This will give you a greater appreciation of the "what" and "why" of accessibility. With this appreciation, you can better define the specific needs in relation to the product you're building, channel your skills in the most effective way, and determine how best to include people with accessibility needs in the product development process so that the product you build is usable, and hopefully enjoyable to use, by everyone.

DOI: 10.1201/9781003288060-4

Many current and potential users of digital products have accessibility needs. Problems occur when digital products haven't been built to accommodate these needs or when products are built on a misunderstanding of user needs. By appreciating the breadth of accessibility needs that exist, you'll be better equipped to define a digital product's accessibility requirements.

3.1 Accessibility Needs versus Disabilities

Throughout this book, we focus on accessibility needs rather than the nature of the disabilities that lead to those needs. As an engineer building digital products for use in a social context, we believe it's less important for you to know in-depth details of conditions that lead to disability than to know about the accessibility needs that arise from disability. That said, a high-level appreciation of conditions that lead to accessibility needs can help to understand the context of accessibility and help design to accommodate those needs.

We also argue that it's of relatively limited value to precisely quantify the number of beneficiaries of accessibility. If you're reading these words, you already know that accessibility is important. Doing accessibility benefits people. Not doing accessibility will mean barriers for some users. It is, though, important to know that the number of beneficiaries is large and fluctuates depending on time and context of use. And we recognize the reality that you may encounter stakeholders who need convincing of the size of the market for accessibility features before providing their support for accessibility efforts. Being an effective accessibility-informed engineer means being able to make a persuasive case for doing and prioritizing accessibility, which may include quantifying the benefits of accessibility efforts. To help with that effort, we provide guidance on where to find relevant data and data sources in Chapter 7, *Accessibility in Practice*.

3.2 Users and Accessibility Needs

To start our journey in describing and categorizing accessibility needs, we can look at ways to think about users and accessibility needs. The following definitions are excerpted from the standard ISO/IEC 29138-1, *Information technology—User interface accessibility—Part 1: User accessibility needs*. They provide a helpful framework for different dimensions: accessibility, assistive technology, user needs, and user accessibility needs.

- **Accessibility:** extent to which products, systems, services, environments, and facilities can be used by people from a population with the widest range of user needs, characteristics, and capabilities to achieve identified goals in identified contexts of use.

- **Assistive Technology:** hardware or software that is added to or incorporated within an ICT system that increases accessibility for an individual.
- **User Need:** prerequisite identified as necessary for a user, or a set of users, to achieve an intended outcome, implied or stated within a specific context of use.

  EXAMPLE 1 A presenter (user) needs to know how much time is left (prerequisite) in order to complete the presentation in time (intended outcome) during a presentation with a fixed time limit (context of use).

  EXAMPLE 2 An account manager (user) needs to know the number of invoices received and their amounts (prerequisite) in order to complete the daily accounting log (intended outcome) as part of monitoring the cash flow (context of use).
- **User Accessibility Need:** user need related to features or attributes that are necessary for a system to be accessible.[1]

People with accessibility needs can be categorized into multiple groups. Some people may be members of more than one group; some people may move from one group to another over time. In this chapter, we focus on the following groups:

- People with disabilities
- People with accessibility needs related to age
- People with temporary impairments
- People with situational impairments

As we discuss each group in more detail, you might realize that, in sum, these groups cover the majority of people who might use your digital product, which makes a pretty compelling argument for accessibility! But before we explore each group in detail, we'll first consider three characteristics of any group of current or potential users of a digital product—ability, aptitude, and attitude.

Ability, aptitude, and attitude were used as significant influencing characteristics on older adults' use of technology in research conducted in the early 2000s by Dana Chisnell and Ginny Redish.[2] They have become three user characteristics that can help distinguish members of any group—regardless of age or disability—and put their experience into the context of a wider audience.

- **Ability:** Some people may have severe disabilities that significantly restrict sensory, physical, or cognitive capabilities. Some may have flare-ups of conditions that happen to coincide with the time of the evaluation. Some may have less severe impairments that, in combination, cause more significant issues. Some may have impairments that do not appear to inhibit their use of technology. And some may have no impairments at all but nevertheless have accessibility needs.

• **Aptitude:** Aptitude covers the related but distinct characteristics of experience (time spent using technology) and expertise (knowledge and skills acquired). Participants' aptitude for using digital products, including their assistive technology, may vary significantly. Some people may have been using digital products for many years; others may have less experience. Some people may have very narrow experience and expertise, and may be very comfortable using some technologies but not others. In general, greater aptitude can help increase general efficiency and improve problem solving and error recovery.

• **Attitude:** Some people will be curious, persistent, and confident, including when using digital products. Others may have limited tolerance for exploring or recovering from errors. Attitude may affect the effort someone makes in attempting a task using a digital product or continuing after encountering a problem. Attitude is a complex product of a range of factors, including health, chemical makeup, and neurology, combined with lived experience and incentive to use a digital product, and a limited attitude should not be dismissed as a simple lack of personal responsibility to persist until successful.

As you become familiar with the categories of people who have accessibility needs, also bear in mind diversity within each category across these three characteristics. Keeping the ability, aptitude, and attitude dimensions in mind will help you avoid making stereotypical assumptions that adversely affect the validity of design and development decisions.

3.3 People with Disabilities

Disability is the first and most significant influencing factor for user accessibility needs. In this section, we focus on the categories of disability and corresponding user accessibility needs that are most relevant to digital product design and implementation. Categorization of disability groups and subgroups can quickly become complex and, in some cases, contentious. Historically, there has been a trend of greater attention to some disabilities at the expense of others, leading to an imbalance of information and understanding about certain conditions and neglect of some groups, particularly relating to cognition. Definitions of disability and terms used for categories and sub-categories can vary depending on location and culture, so where helpful, we'll note different terms used to refer to the same group.

3.3.1 Cognitive Disabilities

Some people have disabilities that affect cognitive functioning, including the ability to read, learn, process, remember, and communicate information in various formats and in different contexts. The group of people with accessibility needs related to cognitive functioning is large and varied, yet it is one whose needs are less well understood and defined in comparison to other disability groups. Some members of this diverse group have historically experienced discrimination, including institutionalization and loss of basic human and civil rights.

As of the time of writing, there is a significant effort to increase recognition of, respect for, and support for diversity relating to cognitive functioning. Along with this effort is an evolution of the language used to describe certain conditions. Many national and international differences still exist in definition of certain terms, so find and use terminology that best describes cognitive accessibility within your context.

W3C WAI's Cognitive and Learning Disabilities Accessibility Task Force (COGA)

In an effort to address the gap in accessibility requirements that address user accessibility needs associated with cognitive accessibility, the World Wide Web Consortium (W3C) Web Accessibility Initiative (WAI) established the Cognitive and Learning Disabilities Accessibility Task Force (COGA). COGA's work has focused on researching, understanding, and documenting user needs, influencing requirements included in the Web Content Accessibility Guidelines that address specific areas of cognitive accessibility, and working to make Web Accessibility Initiative output more cognitively accessible.

For more information, visit: https://www.w3.org/WAI/GL/task-forces/coga/

To help distinguish the different circumstances and user needs within the broad category of cognitive disability, we can consider a number of distinct groups of needs. Some people have conditions that mean they are members of multiple groups.

3.3.1.1 Learning Some people have difficulty with the process of gathering information presented in content in different formats and using that information to acquire knowledge and make decisions. Sometimes referred to as perceptual disability, examples include dyslexia, dyscalculia, and dysgraphia, which respectively refer to impairments of reading, working with numbers, and writing. In some countries, this group of conditions may be referred to as a learning disability, learning disorder, or learning difficulty.

User needs to support learning include the ability to:

- Change the display of content to make it easier to read, including changing typeface, text size, line spacing, and letter spacing.
- Access content presented in formats other than text, such as images, video, and animation.
- Simplify the presentation of primary text content by removing images and other potentially distracting content.
- Access content that is presented in consistent layouts, using consistent terminology, and which behaves in predictable ways.
- Receive support when inputting data into forms and when checking for errors.

3.3.1.2 Memory and Attention   Some people have difficulty remembering information over time. This can occur as a result of a range of conditions that affect short term or working memory. Some people have conditions that make it more difficult concentrating on a specific task and are more susceptible to distraction.

User accessibility needs to support memory and attention include the ability to:

- Access content that is presented in consistent layouts, using consistent terminology, and which behaves in predictable ways.
- Receive support when inputting data into forms and when checking for errors.
- Pause or turn off animations and regularly changing content.
- Simplify the presentation of primary text content by removing images and other potentially distracting content.

3.3.1.3 Communication   Some people have conditions that affect their ability to communicate in text. They may use symbol-based communication in place of or in addition to text.

User accessibility needs to support communication include the ability to translate content into symbol-based language.

3.3.1.4 Intellectual Disabilities   Intellectual disabilities significantly affect a person's capacity to learn, including the rate of learning and the ability to learn more complex subjects, and may also impact the ability to use input devices such as a mouse or keyboard. Intellectual disability may be referred to in some countries as developmental disability or learning disability.

User accessibility needs related to intellectual disabilities include the ability to:

- Access content presented in formats other than text, such as images, video, and animation.
- Access content that is written in simple language and avoids metaphor, analogy, and other abstract ways of presenting information.
- Clearly distinguish and easily activate controls with minimal chance of error.

3.3.1.5 Neurological Conditions   Neurological conditions include seizure disorders such as epilepsy, which can lead to increased risk of a person experiencing a seizure when exposed to content that flickers or flashes at certain frequencies.

User accessibility needs focus on the ability to avoid exposure to flickering or flashing content.

3.3.1.6 Neurodivergency   Neurodivergency is a term that has emerged from a social justice movement rather than from medical origins and is used to describe a growing recognition that there is a broad diversity in the ways that people think, learn, and behave.[3] This approach challenges a historic assumption that there is a "normal" way and that outliers need to be treated, or "corrected," in the way they think, learn, and behave. In some cases, neurodivergent people may be disabled most by attitudes of others in society, rather than by their own condition. Neurodivergent people include people with autism spectrum disorders, including Asperger's syndrome and autism.

User accessibility needs for neurodivergent people focus on minimizing sensory and cognitive overload, reducing demands on short-term memory, and avoiding complex input methods.

Every engineer should know... about neuro-inclusive digital accessibility.

By Lē Silveus

In the ever-evolving world of web development, crafting online experiences that are accessible and user-centered is vital. However, the concept of accessibility extends beyond complying with WCAG guidelines. We're going to dive into an important aspect that often gets overlooked: neurodiversity. As product creators, understanding and embracing neurodiversity can lead to more inclusive and effective digital products while catalyzing innovation.

Neurodiversity refers to the natural variation in the human brain and encompasses individuals with both majority and minority neurological profiles. It is the spectrum of the human mind. It includes people whose way of thinking and being in the world is currently accepted in our society and those who are not. People with labels such as autism, ADHD, dyslexia, and others are referred to as neurodivergent or neurominority because, in recent history (at least the last 150 years), they have been severely marginalized in many nations of the world.

Just as biodiversity is crucial for a thriving ecosystem, neurodiversity is essential for a vibrant and inclusive society. To bring back this neurodiversity and repair the damage done by marginalization, we need to learn to include, love, and appreciate the often-unique skills and perspectives of neurominority people. In tech, this means adding them to the accessibility and Diversity, Equity and Inclusion conversations and bringing them to the creative table.

You play a pivotal role in shaping the digital landscape. By recognizing and accommodating neurodiversity, you ensure that your creations are accessible to a broader range of users. This isn't just about altruism; it's about making practical, user-centered design choices that improve the overall quality of your products.

In the dynamic realm of web development, a commitment to accessibility goes beyond meeting basic guidelines. By embracing neurodiversity, digital creators can create products that cater to a broader spectrum of users.

3.3.2 Physical/Motor Disabilities

Some people have disabilities that affect the ability to move and control movement, in particular the use of the hands to operate a computer, smartphone, tablet, kiosk, or other hardware device. Within this category of disability, there are a range of diverse conditions that cause:

- Changes to fine motor control, leading to tremors, unpredictable movement, or difficulty moving hands and fingers, making it difficult to use a standard mouse, keyboard, or touchscreen, or to operate buttons on a hardware device.
- Reduced sensitivity in fingers, making it more difficult to operate devices, including touchscreens.
- Increased pain when attempting to make physical movements.
- Reduced energy levels when attempting to make physical movements.
- Paralysis and loss of use of one or more limbs.
- Loss of limb.

Some physical disabilities are sufficiently severe that people are unable to use their limbs (hands and/or feet) to control their device and instead use other parts of their body to control input methods, such as speech or eye-gaze.

User accessibility needs related to physical/motor disabilities include the ability to:

- Efficiently interact with a digital product using a keyboard or keyboard-substitute input device.
- Adjust the operation of a keyboard, to reduce the effort required to accurately press a key.
- Visually track the progress of keyboard focus through an interface.

- Efficiently interact with a digital product using speech input.
- Make it easy to accurately hover or click a mouse or other pointing device.
- Make it easy to interact with a touchscreen interface without requiring complex gestures.
- Adjust timeouts to ensure there is enough time to complete actions.

3.3.3 Sensory Disabilities

Sensory disabilities affect the ability to perceive information using specific sensory channels. For digital accessibility, the main sensory areas of focus are vision and hearing.

3.3.3.1 Visual Disabilities   Some people have disabilities that affect the visual channel and are not fully correctable through glasses or contact lenses. The impact may range from loss of functional vision to reduced visual acuity, field of vision, or color perception.

Low vision and sight loss can be caused by many different conditions, including eye disease such as macular degeneration, retinitis pigmentosa, and cataracts, and conditions such as diabetes, albinism, brain injury, and cancer. People with a specific condition may experience different combinations and severity of vision loss. Some people with low vision may experience a gradual loss of sight that may lead to blindness; others may have a specific, stable condition that does not change over time.

People who are blind have little to no functional vision. The primary user accessibility need for people who are blind is information that is available through the audio channel. Some people who are blind use braille or other tactile representations of content, so an additional user accessibility need is the ability to perceive content in tactile form in addition to or as an alternative to auditory output. Many blind people do not read braille, although finding reliable data on the percentage of people who are braille literate is challenging.[4]

Accessibility needs of people who are blind center around the ability to receive all content in non-visual format, most importantly through audio and tactile channels. This includes the ability to access:

- Equivalent alternatives to information provided in images.
- Equivalent alternatives to information provided through visual characteristics such as color, shape, and size of text and objects.
- Information about the structure of page content and relationships between pieces of content, necessary to support understanding the content.

- Information about each interactive user interface control, including its name, type of control, and current state or value.
- Notifications indicating changes to content or presence of errors.

People with low vision have functional vision that is impaired in some way. In some regions, such as the UK, this group may be referred to as partially sighted. Types of vision impairment include:

- Reduced visual acuity, where the ability to distinguish detail is significantly reduced. This may affect near (close) or far (long-distance) vision, and in more significant cases may make it difficult to distinguish text or graphics.
- Reduced field of vision, where someone may have difficulty or be unable to see specific areas of the field of vision. For example, people with the condition often referred to as tunnel vision may only be able to see the center of the field of vision.
- Reduced sensitivity to light, especially bright light.
- Reduced sensitivity to contrast, affecting the ability to distinguish similar hues.

Reduced color vision or color deficit refers to a person's difficulty or inability to distinguish specific pairs of colors. This condition is often also referred to, less accurately, as color blindness. Common instances include:

- Protanomaly and protanopia, respectively, the difficulty and the inability to perceive red light.
- Deuteranomaly and deuteranopia, respectively, the difficulty and the inability to perceive green light.
- Tritanomaly and tritanopia, respectively, the difficulty and the inability to perceive blue light.

Protanomaly, protanopia, deuteranomaly, and deuteranopia can make it difficult or impossible for someone to distinguish red and green hues; red and green colors may appear to have identical hues, such as dark green or brown. Tritanomaly and tritanopia can mean that blues and yellows appear similar or identical. The condition of monochromacy is the inability to perceive any color at all, and compared to conditions that affect the perception of specific colors, it is extremely uncommon.

Accessibility needs of people with low vision and color deficits vary depending on the nature of the impairment. Many accessibility needs of people with low vision and color deficits focus on the ability to change the appearance of an interface, including the ability to perform one or more of the following:

- Magnify or zoom in on all content, including text and images.
- Increase the text size.

- Customize the display of indications of input location, such as keyboard focus, pointers, and cursors.
- Reflow content into a single column in order to make it easier to scan.
- Change the text and background color schemes.
- Change the typeface.
- Change the text and line spacing.
- Reduce the brightness of a screen.
- Change the contrast of hues—for some people, to increase contrast, and for others, to reduce contrast.

Another low vision accessibility and color deficit user need is the ability to perceive information without needing to distinguish hue, brightness, and other visual characteristics of content.

3.3.3.2 Auditory Disabilities

Some people have disabilities that affect the auditory channel and are not fully correctable through a hearing aid. The phrase "hard of hearing" categorizes conditions that cause people to have some hearing loss, which may reduce their ability to distinguish some sounds from others or to hear sounds of a particular volume or frequency. Complete hearing loss is usually referred to as deafness.

People who are deaf may identify as Deaf (with a capital "D"), indicating membership of a culture of Deafness, connected through a shared language, usually a sign-based language. Some people who are deaf, particularly those who were prelingually deaf or lost their hearing during childhood, use sign language as a primary means of communication, and may have lower literacy in text-based languages.

Accessibility needs of people who are deaf or hard of hearing include the ability to:

- Access alternative versions of audio content, for example, through text or sign language interpretation.
- Adjust the display of text alternatives for audio.
- Access sign language versions of text content.
- Adjust the volume level of audio content.
- Use headphones to access audio content rather than open audio.

It's worth noting that sign languages use distinct syntax, grammar, and vocabularies and have evolved in different patterns compared to spoken language. So don't assume that because two countries may share a common spoken or written language, their sign languages are also mutually intelligible. For example, American Sign Language (ASL) is more closely related to French Sign Language (LSF) than British Sign Language (BSL). This means that serving a multilingual audience may require sourcing and providing content in multiple sign languages, just as you would need to consider a strategy for internationalization and localization of text and spoken content.

3.3.3.3 *Combined Visual and Auditory Disabilities* Some people have a combination of limited or no functional vision and limited or no hearing, or deaf-blindness. For deafblind people, the ability to receive meaningful information in tactile format is an essential user accessibility need. This may be provided through hardware that converts digital content to tactile output or by tactile representations of content, such as tactile diagrams.

3.3.4 Speech Disabilities

Some people have disabilities that affect speech, specifically the ability to produce speech that other people (and machines) can understand. This includes conditions that may cause:

- Stuttering.
- Involuntary utterances, such as Tourette's syndrome.
- Changes to the muscles required to produce speech, such as cerebral palsy.
- Difficulty or inability to utter specific sounds.
- Changes to the ability to vocalize words in the correct order.

Some conditions, such as certain brain injuries, anxiety, or hearing loss, may result in a temporary or long-term loss of speech.

People with speech impairments have accessibility needs for any interface that relies on speech as an input device, including the ability to:

- Control the interface and input data through methods other than speech.
- Specify communication preferences that are alternatives to voice-based communication.

3.3.5 Chronic Illness

Some people have chronic illnesses, such as cancer, multiple sclerosis and endometriosis, which limit the amount of physical and mental energy they may have in a given day. The impact of chronic illness may vary within and between days and is likely to be unpredictable, meaning that someone may be unsure as to their energy capability on any given day. In addition to conditions that affect sensory perception, motor skills, and cognition, people with chronic illnesses may experience higher levels of pain and fatigue, which can limit their ability to interact with digital products for a prolonged time. People with chronic illnesses may be particularly adversely affected by the challenge of complex and inefficient interfaces combined with the cumulative impact of multiple accessibility barriers experienced during interaction. Despite the growing incidence of chronic illness, and the fact that people with chronic illnesses are covered by disability rights legislation in many countries, it's often overlooked or given less attention in comparison to other disabilities.

Mental health conditions such as post-traumatic stress disorder (PTSD), depression, and anxiety can often be considered chronic. Often triggered by specific stimuli or environments, these conditions can create a range of user accessibility needs, in particular at times where stress is heightened.

User accessibility needs for people with chronic illnesses focus on minimizing complexity, redundancy, and inefficiency in the design and behavior of digital products and providing ways to reduce the effort required to navigate, perform tasks, and recover from errors.

3.4 Aging and Disability

As life expectancy increases, so does the number of older technology users, raising the significance of older people as beneficiaries of accessible digital interfaces. The relationship between aging and disability is multifaceted, and many older adults may not consider themselves disabled. The aging process brings changes in a range of abilities, including vision, hearing, fine motor skills, and short-term memory, and can exacerbate the nature of any existing disabilities. These abilities will generally decline with age, but with great individual variety in terms of timing, combinations, and rates of change.

So, for example, within a sample of people aged 75, not everyone will have the same user accessibility needs. In that group, there will be a range of needs that may change with time, and some users may not be familiar with those needs or how to address them. What is certain is that, within any sample of older adults, a significant proportion will benefit from accessible product design.

It's also worth taking a moment to address a common stereotype of older adults being unable or unwilling to use digital products. The dimensions of ability, attitude, and aptitude we introduced earlier in this chapter are important reminders that in a sample of older adults, there will be people who have been using the web for more than 25 years, people who have had a smartphone since the early 2000s, as well as people who are less confident and experienced using websites and may never have owned or used a smartphone. Not all older adults are uncomfortable with digital technology, just as some younger people may have limited ability, attitude, and aptitude for using digital products.

On that note, at the other end of the age spectrum, young people can also benefit from accessible design. In particular, younger children share many of the user accessibility needs addressed in the range of cognitive disabilities.

Every engineer should know... people with disabilities have a wide and diverse range of user needs.

By Jonathan Avila

It is essential for people who design, create, and validate technology to understand the needs of people with disabilities, get feedback from users with disabilities, and understand the technology and settings people with disabilities use to access various types of experiences. You should engage diverse users, including disabled people, in user research, testing, and feedback sessions to gain a better understanding of their needs, preferences, and challenges. This involvement helps to create digital products that are truly accessible and meet the unique requirements of these users.

Understanding the unique user needs of people with disabilities allows for the development of more inclusive and user-friendly interfaces that cater to a diverse range of users. Knowing what works well and what is a problem for different people is necessary to avoid creating experiences that can exclude. For example, one might assume the needs of users with low vision are limited to things like contrast and magnification, but what is the impact of a limited field of view caused by a person looking closely at the screen, for example, from a distance of 4 inches or less, or when magnification is used, allowing view of 1/16th of the screen at a time, or when the person has visual field loss, limiting what is seen at any given time? In this case, the impact of the proximity of controls and their labels, changes in focus, and new content that appears on the screen suddenly becomes apparent when you understand the way a person with low vision experiences your interface.

3.5 People with Temporary or Situational Impairments

The terms temporary impairment and situational impairment are used to describe impairments that are not classified as disabilities but that could be considered disabling. The source of a situational impairment is environmental, whereas temporary impairments are physiological, caused by injury, illness, or other conditions that change a body's capability for a short period of time. The user accessibility needs associated with temporary and situational impairments are like those related to disability, even if the circumstances may be very different.

3.5.1 Temporary Impairments

People may experience temporary reductions in capability due to injury, illness, or other changes in health that affect how they experience and use technology, including:

- **Injury:** Injuries can affect people's ability to interact with technology. For example, arm, wrist, and hand injuries, including wrist sprains and fractures, can affect fine motor skills. So too can repetitive strain injury, ironically often due to excessive use of input devices for a computer, smartphone, or gaming console.
- **Illness:** Distinct from the chronic illness mentioned earlier, some medical conditions can be sufficiently debilitating and last for sufficient periods of time to create temporary accessibility needs. For example, vestibular disorders such as labyrinthitis affect the inner ear and, consequently, the body's sense of balance. Symptoms include dizziness and vertigo, which may make it difficult to focus on a screen for more than a short time, and user interface animations may cause nausea.
- **Changes to Capabilities:** Temporary changes to physical abilities can impact technology use. For example, temporary speech impairments, such as numbness of the mouth following dental treatment or slurred speech due to reactions to medication, could lead to someone encountering difficulties operating a speech-enabled interface, especially where the interface has been trained to recognize their voice under specific conditions. Cognitive abilities can be affected by a range of situations, such as sleepiness due to jetlag or the effects of medication. These situations can lead to a reduced ability to retain attention or process information.

3.5.2 Situational Impairments

People may experience a temporary reduction in capacity due to environmental factors such as location, weather, technical constraints, and stress level that affect how they experience and use technology, including:

- **Weather:** Weather can create physical and sensory impairments, most often for interfaces used outdoors. Sunlight can create glare, making screens more difficult to read. Increased perspiration may reduce touchscreen sensitivity, as can rain. Cold weather may mean people wear more clothes, including gloves and hats, reducing dexterity, mobility, and possibly hearing.
- **Location:** Location can create cognitive impairments, for example, when users are unfamiliar with the language of an interface or the idioms, metaphors, and stereotypes that are specific to some regions or cultures. Interfaces of kiosks and public access terminals in locations like airports and railway stations are likely to be used by audiences of different nationalities, many of whom will be unfamiliar with local conventions. Location may also create temporary physical or sensory impairments. For example, using a smartphone while on a crowded bus traveling along a bumpy street may reduce a user's

manual dexterity. Watching a video in a noisy bar may make it difficult or impossible to hear the video soundtrack.

- **Technical:** Technical constraints can lead to situational impairment. For example, restricted data connectivity may cause a user to experience an intermittent or unreliable data connection (which may also be related to location—for example, using an app while on a car or train journey through a mountainous area). Some people may need to restrict data use for financial reasons. Similarly, some people may be using an older device with limited storage and processor power; some people may be using a device running an older version of an operating system or browser, or that lacks necessary software.

- **Stress:** Some digital products are used in situations where a user may be experiencing a higher level of stress, which can affect cognitive, physical, and sensory capabilities. Episodes of increased stress may be experienced by people with certain chronic conditions, but stress can also be related to the degree of severity of making a mistake when using the system or the distress a user might be experiencing at the time of use. Examples might include systems used for immigration processes at border facilities, systems used to collect or present health information such as test results, systems for processing financial transactions like loan applications, or systems used to administer high-stakes tests in an educational setting such as a university. Even the nature and purpose of a digital product can create a higher degree of frustration or stress—think of a family trying to use a travel website to book a discounted vacation during the school holiday period while the price is still affordable, their specific hotel dates are available, and flights are available from their local airport.

3.6 Dynamic and Intersectional User Accessibility Needs

It's easy to assume that user accessibility needs are specific and stable. In reality, a person's accessibility needs are likely to change over time and exist in combination with other needs that influence how best these needs are addressed.

3.6.1 Dynamic Needs

Many disabled people have multiple disabilities, and for many people, accessibility needs are likely to vary over time. The impact of a disability, or combination of disabilities, may change within a day or over the long term. Some conditions may progressively increase in severity, others may stabilize, and others may reduce in severity. A specific condition may start with a single disability and evolve to include others. A person may suddenly acquire a disability as a result of illness or injury and go quickly from having minimal accessibility needs to having significant needs. In parallel, temporary and situational impairments are by definition not permanent and are also subject to change.

So what does this mean for engineers? It doesn't reduce the need for accessible design. But it does mean that you shouldn't consider users with accessibility needs as having a fixed profile. Accessibility needs are dynamic, changing based on different factors and contexts. And the cumulative effect of the combination of minor impairments can create more significant usability challenges than you might expect by looking at each impairment in isolation. Consider, for example:

- The impact of reduced visual acuity plus a minor tremor on a person's ability to accurately manipulate a mouse pointer or touch interface.
- A person at their laptop, trying to book last-minute flights for an urgent trip, typing with one hand, while their other hand holds a cellphone to their ear, discussing options with their travel companion.

We refer to this as "designing for dynamic diversity", a term coined by researchers Peter Gregor, Alan Newell, and Mary Zajicek[5] when talking about effective ways to design for older adults. This approach encourages digital product creators to:

- Recognize diversity in ability within an individual over time and between members of a particular group of users, such as older adults or people with low vision.
- Involve people with accessibility needs throughout the design and development process.
- Provide flexibility in how user accessibility needs are addressed in the design of a digital product, including clearly signposting to users how they may adapt the interface to suit their needs.

3.6.2 Intersectional Needs

Accessibility is about inclusion of people with disabilities, and we already alluded to the impact of the intersection of multiple disabilities on user accessibility needs. We must also recognize that disability intersects with other demographic characteristics. In other words, disabled people belong to other social groups, some of which may also experience discrimination based on characteristics such as race, ethnicity, sex, gender identity, sexuality, economic status, geographic location, education level, and religion. The incidence of certain disabilities increases for people in marginalized groups, sometimes as a result of discrimination that has led to greater exposure to conditions that increase risks to health, and to reduced access to quality healthcare.

One consequence of intersectionality is that accessibility may not benefit all people equally. Accessibility solutions that assume a user will have an expensive smartphone or high-bandwidth data connection may be exclusionary to low-income populations. Disability statistics show that disabled people have lower earnings than nondisabled people and higher poverty rates.[6] This is not to say that accessible design can only ever be partially successful or that you should give up on trying to build accessible smartphone apps. Instead, you may need to take additional factors into account when considering solutions to broader accessibility challenges. A solid and inclusive understanding of your product's target audience will help you make informed decisions.

Given how humans are all individual and unique, it follows that accessibility needs are diverse and multifaceted, with many influencing factors. The good news for designing digital products that address user accessibility needs is that the digital environment has inherent features and affordances that enable people to adapt products to address their individual needs. Recognizing those attributes and designing and building in alignment with the environment is key to creating accessible digital products. One critical task is ensuring products are compatible with the range of software, hardware, and accessibility strategies used by people who have accessibility needs. This starts with understanding what tools and technologies people with accessibility needs may already have at their disposal, so that you can take advantage of that technology. We cover these in the next chapter, *Assistive Technology*.

Takeaways

As an engineer, you should:

- Recognize your responsibility to meet user needs in general and prioritize user accessibility needs to avoid creating barriers and causing harm.
- Have a broad and inclusive perspective when conceptualizing the end users of any product that you design and develop—one that includes people with a range of conditions and circumstances that affect how they experience and interact with technology.
- Appreciate the dynamic nature of user accessibility needs; they arise from many circumstances, change over time, and are experienced on some level by every person.
- Account for the accessibility needs of people with disabilities in product design and development.

Notes

1 ISO/IEC 29138-1, *Information technology—User interface accessibility—Part 1: User accessibility needs.* www.iso.org/standard/71953.html
2 D. Chisnell and J. Redish (2005) *Designing Web Sites for Older Adults: Expert Review of Usability for Older Adults at 50 Web Sites.* Washington, DC: AARP. Available at: http://assets.aarp.org/www.aarp.org_/articles/research/oww/AARP-50Sites.pdf
3 N. Baumer and J. Frueh (2021) *What Is Neurodiversity?* Harvard Health Publishing, Boston, MA, USA. www.health.harvard.edu/blog/what-is-neurodiversity-202111232645
4 R. Sheffield, F. D'Andrea, and S. Chatfield (2022) How Many Braille Readers? Policy, Politics, and Perception. *J Visual Impair Blindness*, vol. 116, no. 1. journals.sagepub.com/doi/full/10.1177/0145482X211071125
5 P. Gregor, A. Newell, and M. Zajicek (2002) Designing for Dynamic Diversity—Interfaces for Older People. *Proceedings of ASSETS 2002*, July 8–10, 2002 Edinburgh, Scotland.
6 *Annual Disability Statistics Compendium.* Institute on Disability, University of New Hampshire, Concord, NH, USA. disabilitycompendium.org

4

ASSISTIVE TECHNOLOGY

Objectives

In this chapter, we explore software and hardware that people who have accessibility needs may use to independently access and operate digital products, as well as ways to design digital products to work effectively with these technologies. We review assistive technology tools and adaptations that minimize or eliminate the disabling effects of sensory, physical, cognitive, and other impairments when using digital technology.

Once you're through this chapter, you should:

- Understand ways that people with disabilities use assistive technology to support accessibility needs.
- Appreciate the range of different ways people use assistive technologies, adaptations, and accessibility strategies to work with technology.

Introduction

We've explored the range of user accessibility needs that exist, the circumstances that lead to those needs, and how a person's needs may change over time. The next step is to understand what solutions are available to support those user accessibility needs in the form of a category of hardware and software referred to as "assistive technology," or AT. Like wheelchairs and scooters in the physical environment, in the digital environment, many people think primarily of screen readers as assistive technology. But there's a wide range of hardware and software that people with disabilities use to efficiently and effectively access and interact with

DOI: 10.1201/9781003288060-5

digital products. As an engineer, your role is to ensure the products you build interact effectively with users' assistive technology.

4.1 Technology versus Assistive Technology

The approach to providing users with assistive technology has changed over time. In years past, AT existed as a complex, niche, expensive technology that was considered to be of use to only a small market. It was often provided through health insurance or by public healthcare and therefore treated more as a clinical product than a consumer product. AT was considered a separate category of technology—an expensive add-on to devices used in the workplace, in schools, in public places, and at home, and not necessarily designed with the user experience in mind.

As technology evolved from the school and workplace to become more pervasive, there's been a trend toward integration of assistive technology into operating systems as core features rather than specialist add-ons. While the primary beneficiaries of assistive technology are people with disabilities, AT is increasingly marketed as mainstream, with providers highlighting ways it makes a device easier to use. This approach aligns with the multiple groups of beneficiaries of accessibility and the corresponding range of user accessibility needs, as we discussed in the previous chapter. There are many everyday technology features that began life as AT intended to address specific disability-related user accessibility needs, including captioned TV shows and movies, voice assistant software, accessibility features such as "dark mode" and "reading mode" display settings, and predictive text functionality. The popularity and utility of these features has become much more widespread.

Today's assistive technology ranges from specialist hardware and software to accessibility settings provided by an application, operating system, or hardware. With growing awareness of user accessibility needs, there has been a trend toward a greater range of accessibility settings provided at the operating system level of a device. Operating systems such as Windows, macOS, iOS, and Android all now provide a diverse set of accessibility options that can be applied across applications running on the device. Even so, for some people, accessibility needs require additional dedicated AT software or hardware products.

Assistive technology varies depending on the user accessibility needs it is intended to address. It may operate as a way to transform the output modality of digital content, alter the presentation of digital content, or provide alternative ways to operate and interact with digital content. Often, a single AT may meet the needs of diverse disability groups.

In her book, *What Can a Body Do?*, Sara Hendren argues that any technology that helps us is assistive. Some technology may not be designed or marketed as assistive technology, yet it may be particularly powerful in helping make digital products easier to use by disabled people. And some assistive technologies help people who may not identify as disabled but may experience situational or temporary impairments. "Every human creature's possibilities are produced, in part, by what's concretely present in its time and place. A body—any body—will take its cues, bend the available resources, and invest its being with the matter around it."[1] This viewpoint supports the social model of disability discussed in Chapter 2, *Disability and Digital Inclusion*, and underpins the approach and guidance in this book. We as designers and engineers of the digital world influence what possibilities people experience in their lives by how we build digital products. This means understanding and building for the ways people use technology and techniques to adapt and extend their capabilities for access to content and functionality.

4.2 Content and Output

There is a range of assistive technologies that support user accessibility needs relating to access to content, including providing alternative ways to output content and adapting content to better address user accessibility needs. These are mainly software-based solutions provided through dedicated applications and operating system settings and adaptations; hardware devices also exist to support access to content.

4.2.1 *Text-to-Speech*

For people with accessibility needs relating to significant sight loss, assistive technology can provide an alternative to the visual channel. Perhaps the most widely recognized example is screen reader software, which provides text-to-speech output of content along with a powerful set of keyboard commands to allow for flexible navigation through content. Examples of standalone screen reading software include:

- For Windows, JAWS (a commercial product) and NVDA (an open-source product) are standalone screen reading applications. Microsoft provides Narrator as a screen reading application that comes as standard with Windows, although at the time of writing, Narrator provides more limited capability than JAWS and NVDA.
- For macOS, iOS, and iPadOS, Apple provides VoiceOver as a standard feature of the operating system.
- For Android, Google provides TalkBack as a standard feature of the operating system.
- On Linux, Orca is an open-source screen reader.
- A version of JAWS is also available to run on Windows- and Android-based kiosk platforms.

WebAIM, a non-profit accessibility organization based at Utah State University, regularly runs a Screen Reader User Survey, reporting on usage statistics of screen readers across different platforms, along with a range of additional useful data on behaviors and preferences. These surveys, conducted approximately every two years, have provided fascinating insights into trends in screen reader use over time. Their 2021 survey reported JAWS, NVDA, and VoiceOver as the three most commonly used screen readers.[2]

In addition to dedicated screen reading applications, there is a category of application that includes text-to-speech functionality, among other features. For example, reading assistance software supports people who have functional vision but have accessibility needs that require assistance in reading and comprehending text content. For example, someone with dyslexia may prefer to highlight a block of text and have it read aloud to reduce the effort required to visually process the text.

Non-visual interaction with digital applications involves a very different interaction paradigm to visual exploration of a user interface and its content. We discuss this critical aspect of digital accessibility in more detail in Chapter 5, *Core Attributes*.

4.2.2 Tactile Output

Some assistive technologies that provide an alternative to the visual channel do so by providing tactile output. These may be used alongside or instead of a screen reader. Tactile printouts or models of content can provide accessible versions of graphic content such as diagrams and maps. Tactile labels and controls on a kiosk or other hardware device can make non-visual operation easier, as can haptic cues like vibration, for example, on watches and smartphones. For people who are deafblind, some form of tactile interaction and output is necessary to work with digital products.

As a more powerful and flexible solution, refreshable braille displays are hardware devices controlled by screen reader software to present screen content as physical braille characters. They rapidly refresh to support quick reading of screen content and vary in size depending on how many characters they can display at one time. Larger braille displays can present up to 80 characters at a time, but have limited portability, so tend to be attached to a standard keyboard for a PC or laptop. Smaller braille displays present fewer characters at one time but are portable and may be linked to a smartphone or other device using a wireless connection. Some braille displays also allow users to create content, using custom keypads to quickly generate braille content. For people who use braille, these devices can significantly improve efficiency of reading and writing.

4.2.3 Magnification

A range of assistive technology is available to customize an interface's visual display to make it easier for people to perceive content. Screen magnification software can magnify the content of the screen to many times the default text size, giving users power over levels of zoom and the means to control them. Since zooming reduces the amount of content shown on screen at any given time, screen magnification software also has functionality to support orientation and tracking around a screen, helping readers maintain focus, locate content, and improve reading performance. Screen magnification software may also include other display customization options and text-to-speech capabilities.

On tablets and mobile devices, functionality to zoom or magnify screen content or change text size is embedded with the operating system and is widely used to make content easier to read and also to address the limitations of a smaller screen. Hardware solutions also exist for magnifying content, for example, through use of a document camera or magnifying glass.

4.2.4 Display Modifications

Assistive technology features to support reading difficulties or reduced vision include hardware, software, and operating system settings that allow users to:

- Set a large text size and reflow content and functionality to accommodate larger text.
- Change color schemes, for example, inverting colors so that content is shown in light text on a dark background or applying a high-contrast color palette to increase readability.
- Adjust the appearance of the focus indicator.
- Change the appearance of the mouse pointer to make it easier to track.
- Reduce brightness and color contrast to reduce eye strain and stimulation.

Note that along with AT to augment the visual channel, like magnification and display adjustments, some people with vision or cognitive accessibility needs may use a screen reader to read out specific screen content. This means you should avoid design decisions that assume all screen reader users cannot see the screen.

4.3 Operation and Input

Assistive technologies that support user accessibility needs relating to operating a user interface include both hardware and software solutions. Alternative input devices are hardware devices with supporting software that together perform a similar role to a standard mouse and keyboard but do not rely on fine manual dexterity for operation. Some alternative input devices work in combination with a dedicated on-screen keyboard to support more efficient input. Some alternative input device options are built into standard operating system accessibility settings, making use of the device's camera or microphone. External keyboards can be wirelessly connected to touchscreen devices such as smartphones, tablets, and kiosks, allowing keyboard operation instead of touch.

4.3.1 Switch Controls

A switch device is a general term for any device that has a limited set of states and can be operated in different ways. For example, sip-and-puff devices act as a switch controlled by the mouth. Other switch devices may be operated by a hand or foot or by moving the head. On screen, the switch device may allow a user to choose between moving focus from one control to another or activating the currently focused control. Alternatively, focus may move automatically in a logical, slow progression through the interface.

4.3.2 Gestures and Movements

Gesture recognition devices involve body-worn sensors that allow specific movements to be interpreted as input actions. Gaze recognition uses a camera to detect gaze direction and manipulate the mouse pointer accordingly. A head wand is a hardware device attached to a headband, allowing someone to use the wand to press keys by moving their head.

4.3.3 Speech

Speech recognition allows a user to issue verbal commands to control the interface, allowing navigation, operation, and data input without needing to use a keyboard or mouse. In recent years, the quality of speech input software has significantly improved, including greater tolerance for speech of diverse accents and volumes. It's important to distinguish between two distinct categories of speech input software, with the first one being more accurately described as an assistive technology supporting user operation of an interface:

1. Dedicated speech input software intended to enable a user to operate a user interface by, for example, selecting a specific hyperlink and activating it. This software typically responds to a set of specific, terse commands.
2. Voice assistance software, such as Apple's Siri or Microsoft's Cortana, whose primary purpose is to answer questions expressed in a more natural language by interrogating large repositories of data.

4.3.4 Input Modifications

Some user accessibility needs may be supported through hardware or software adaptations to standard input methods. These can help accommodate tremors and other conditions that affect fine motor control, as well as reduce the physical effort required to operate an input device. Hardware modifications focus on adjusting hardware design to reduce the physical strain of operation, for example, through ergonomic keyboards, trackballs, and joysticks.

Some software modifications focus on adapting the signals received from a hardware input device to make operation easier and reduce input errors, such as adjusting the click rate of a mouse button and touchscreen sensitivity to account for involuntary clicks and taps.

Other software modifications focus on adjusting the visual display of the pointer, keyboard focus, or cursor to help meet user accessibility needs relating to vision and motor control. For example, operating system and software features allow users to customize the size and appearance of the mouse pointer or cursor, including adding a trailing animation when the pointer is moved.

Most alternative input solutions operate as input devices in the same way as using the TAB and ENTER keys on a standard keyboard, where TAB supports linear progression through each user interface element in turn and ENTER activates the currently focused element. This means that designing a digital product that can be navigated and operated by a keyboard increases the chances that the product can also be operated using alternative input methods.

4.4 Open and Closed Functionality Systems

Not all systems are equal in their capability to provide assistive technology to suit an individual set of user accessibility needs. For personal or workplace devices like laptops, smartphones, and tablets, it's very likely that a user has sufficient flexibility and rights to install, manage, and customize AT to support accessibility needs. We refer to such devices as open functionality systems, where your aim as an engineer is to support users' assistive technology, not to build it in. Web and software applications are designed to work with open functionality systems, which means the primary goal of accessibility efforts is to support interoperability with a broad range of AT.

However, some systems have much less flexibility for users to install and configure assistive technology to meet their specific needs, such as public access terminals, kiosks, timeclocks, and point-of-sale machines. The function of these devices means they are designed for use by a large number of people, with each individual usage session often limited to a short transaction. They are likely locked down for security reasons. This means it may be difficult or impossible for someone with accessibility needs to make adjustments, such as changing accessibility settings, using screen reading software set to their own preferences, or connecting an alternative input device. Other devices with specialized or limited hardware and software capability, from gaming consoles to household appliances with user interfaces, may also have restricted capacity to provide flexibility in input and output methods and settings. We refer to these as closed functionality systems, and the aim with these systems is to build in support for user accessibility needs.

As an engineer, the implications of the restrictions of a closed functionality system are that any application running on the system must be operable and accessible out-of-the-box. This in turn means it's necessary to provide accessibility features and adaptations as part of the hardware and software interface to the best of its capabilities. If you are building a digital product that runs on a closed system, you'll need to provide assistive technology options in a way that you wouldn't need to on an open functionality system, including providing users with the necessary guidance on how to access and use the available assistive technology. We'll discuss accessibility for closed systems more in Chapter 8, *Requirements Specification*.

4.5 Using Assistive Technology

As an engineer building accessible products, you should know about the variety of assistive technologies available to meet different user accessibility needs and the different ways in which you should use them. If you have a disability, you may already be very familiar with and adept with some assistive technologies. But for engineers who lack experience using an assistive technology or observing people using AT, it may be difficult to appreciate how the technology works, let alone how to use it in a competent manner. How familiar an engineer should become with AT is an interesting question, and our view is that there are two answers.

First, any assistive technology that is not part of your standard toolkit should be a target of research and study. Learn as much as you can about how these software and hardware tools are used by people who use them every day. Ask family, friends, and colleagues who use AT to share their strategies and demonstrate their techniques. Seek out videos of people with disabilities using AT for tasks using their smartphone, laptop, or gaming console. Go to conferences, classes, and training sessions where you can interact with people using AT and ask questions. With this approach, you're not seeking to become an expert. Instead, you're learning about the diversity of assistive technologies from the experts who use them, building up an understanding of the accessible user experience and what you can do as an engineer to optimize that user experience.

Second, focus on learning enough about using key assistive technologies to support accessibility testing. Here, you're seeking to use assistive technology as part of accessibility test procedures to verify whether the interface behaves as intended and whether actual behavior matches expected behavior. For example, knowing commonly used keystrokes and swipe gestures for different screen reader navigation options can help when testing functional code for potential accessibility issues. Be aware, though, of differences between versions of the same AT product and differences related to how well an assistive technology product works with other software, especially browsers.

Given the rate of change and innovation in the digital environment, no one can achieve full expertise with every assistive technology in every context and use case. With the exception of the AT that you use in your daily life, gaining expertise in every AT and modification is probably not worth the investment if your goal is to engineer accessible digital products. But being curious about all the different ways people use technology and asking experts to share their perspectives is definitely worthwhile, as is understanding the range of software, hardware, and techniques. What *is* essential to engineering accessibility is respecting the inherent affordances and features that make the digital environment work in harmony with assistive technology. We discuss these core attributes in the next chapter.

Every engineer should know… people with disabilities use a variety of assistive technologies and accessibility strategies.

By Jonathan Avila

Engineers should know how assistive technologies operate and how users with disabilities interact with them. People with disabilities may use various techniques, accessibility features, and assistive technologies throughout the day and may switch to different ways of using technology in different contexts. Familiarize yourself with the various assistive technologies, including screen readers, screen magnifiers, speech recognition software, reading assistants, and alternative input devices like switch control systems. Understand how accessibility features such as browser zoom, document outlines, reduction in animation, contrast modes, and typography settings are used.

Without proper understanding, one may believe assistive technology or accessibility features are working as expected when they are not, or they may believe they are not working correctly with their product when they are working correctly. In addition, learning and practicing how people with disabilities use the technology will allow for correct usage and feedback. For example, screen readers are designed to be used with a keyboard, as the demographic of users who are blind are keyboard-only users, and thus commands may only work as intended when used this way. While some non-screen reader users may assume screen reader users access a webpage by listening to the whole page in one go, instead native screen reader users generally navigate through structures such as headings and links, reading blocks of content, searching for specific items, and tabbing through form fields.

Familiarity with assistive technology is also important because, without experience, you may think the technology is too difficult or complicated and get the impression that a person with a disability cannot perform a task because you do not know how to do it using the technology. Using assistive technology can also provide an understanding of the obstacles faced by people with disabilities when technology is not designed to be inclusive, when you directly experience these barriers with the technology.

Also, not all people with disabilities use assistive technology; some use accessibility features, and others may use standard features in a way that you did not consider, such as use of browser zoom, which triggers responsive web variations on the desktop that engineers only assumed would be present on small screen devices like mobile. In cases like the above, it may have been assumed that responsive menus don't have to be keyboard accessible because they will only appear on mobile. This assumption is not accurate—not just because of how responsive variations may be accessed on the desktop but also based on how users with physical disabilities may use mobile devices.

Takeaways

As an engineer, you should:

- Appreciate that the intent of all technologies is to assist humans in accomplishing tasks and the role of assistive technologies in extending and adapting technologies to work for people in different contexts.
- Consider the myriad ways people will use AT to interact with the products you design and develop and take into account those experiences and use cases.
- Understand and account for the different accessibility support requirements for open and closed functionality digital products.
- Become familiar with using different assistive technologies to inform and facilitate design, development, and evaluation.
- Seek out perspectives from expert AT users when designing and building digital products.

Notes

1 S. Hendren (2020) *What Can a Body Do? How We Meet the Built World.* United States: Penguin Publishing Group.
2 WebAIM (2021) *Screen Reader User Survey #9 Results.* webaim.org/projects/screenreadersurvey9

5

CORE ATTRIBUTES

Objectives

In this chapter, we cover the essential features and attributes of an accessible digital environment. We discuss how these features influence approaches to ensuring accessibility of content and interaction, particularly for people using assistive technologies.

Once you're through this chapter, you should:

- Understand and appreciate foundational digital accessibility concepts.
- Recognize how accessibility features and attributes of the digital environment support accessibility and assistive technologies.
- Be ready to support these attributes in your design and engineering practices.

Introduction

When we introduced digital accessibility in Chapter 1, we considered the parallels between accessibility in the physical world and accessibility in the digital world, and some of the attributes that exist in an accessible physical environment. Here we introduce four core attributes that support accessibility in the digital environment: compatibility with assistive technology, support for linearized access to content and keyboard interaction, flexibility and support for adaptation, and programmatic access to accessibility information. As an engineer, it's essential that you adopt these attributes as core requirements when building digital products. They will reappear throughout the book as we cover the practicalities of defining accessibility

DOI: 10.1201/9781003288060-6

requirements and applying the process of designing, implementing, testing, and managing digital products to ensure they satisfy these requirements.

5.1 Compatibility

When approaching the task of building an accessible digital product, there's no need to reinvent technology that already exists. If you're an assistive technology user, you may have experienced well-intentioned but ineffective attempts to replicate the functionality of your assistive technology within an application or website. Unless you are developing AT software or hardware, in most cases, you will not need to provide AT functionality in the digital products that you build, particularly with open systems. In these cases, your task is to build digital products that are *compatible* with AT, creating a partnership where your digital product provides the information an AT needs to be able to operate successfully.

5.1.1 Device Independence

As we learned in the previous chapter, people with accessibility needs may use a diverse range of hardware and software to support interaction with digital products. People may also make a wide range of personalized adjustments to software and hardware settings in order to access and receive content and interact with functionality. Device-independent design focuses on supporting diversity of input and output devices as far as possible rather than assuming that, for example, all users can see the interface and are using a standard mouse and keyboard. Designing for device independence means careful research into the diverse ways in which people might be expected to use a digital product and using this information to inform design, development, and content creation. When you can avoid making engineering decisions that are based on inaccurate assumptions about things like a user's screen size, input device, or ability to perceive user interface objects in a specific way, you reduce the chances that you build a digital resource that only works for some people in some situations.

5.1.2 Interoperability

Interoperability relates to the ability of two or more systems to work seamlessly together with minimum friction for a user. In an accessibility context, this means designing and developing digital products that work effectively with assistive technology, whether that's a dedicated software application or accessibility settings adjustments made at the operating system level. "Working effectively" means ensuring that AT can do its job, whether that's as an input device, routing output to a specific sensory channel, or adjusting how that output is presented in a particular modality. Assistive technology depends on digital products to expose the information needed to enable users to receive the relevant input or output support.

The need for knowledge of front-end development

In back-end systems engineering, accessibility needs to be factored into any effort to develop data structures and other system representations of users and their characteristics. But the majority of the focus of accessibility efforts will be on the front-end—the point where users interact with the system. Depending on your education, interests, and experience as a software engineer, you may have limited experience with front-end code, whether for the web, native mobile apps, or other software development platforms. In practice, especially for large and complex systems, front-end code is often an afterthought, added on to support access to the underlying system architecture and integrations that command so much attention. This is one cause of software interfaces that are difficult to use for everyone and impossible for some. For example, if interoperability isn't considered at the start, extracting the necessary accessibility information from front-end code to allow users to perceive and operate the interface with their assistive technology will be difficult, never mind making that experience usable for assistive technology users.

Take the time to learn about front-end design and development. Your efforts on system architecture, modeling, and engineering will be wasted if the system created isn't usable by people.

5.2 Linearization and Keyboard Interaction

Some people rely on or prefer to use screen reading software to interact with a digital device and the applications that run on the device. An understanding of the screen reader experience can help you prioritize design decisions that help rather than hinder screen reader users—decisions that can benefit other users, too. While screen readers may provide some visual feedback for users who have some vision, their functionality is designed primarily for nonvisual access, supporting people who rely on speech and, in some cases, braille for output when interacting with an application and its content. There are two key characteristics of the screen reader user experience: linearized experience of content and keyboard-driven navigation and orientation. These attributes also impact the experience for other assistive technology users and accessibility strategies.

5.2.1 Linearized Experience of Content

Early computing system interfaces supported a text-based linearized presentation of content, requiring typing commands to control the application. This type of interface is still present in command-line or console interfaces and other user interfaces that do not support graphical content presentation. However, the advent of the graphical user interface in the early 1970s created today's familiar paradigm of two-dimensional layout of content and controls, sometimes with layers of content, such as drop-down menus and dialogs. This two-dimensional layout enabled rapid visual scanning of content and available functionality. It also provided new ways for interface designers to use visual properties to group and distinguish interface components. For many users, a visual, multi-dimensional graphical user interface may seem like the only possible way to interact with a digital product.

By contrast, the screen reader user experience is linear in nature, regardless of whether the visual interface is text-based or graphical. This is because the screen reader gathers all elements of the user interface and outputs them in linear order as speech. If you're not familiar with how a screen reader presents digital content and supports interaction, imagine a radio news bulletin. In this bulletin, structured content is presented in a particular sequence—the title of the bulletin, the headlines, then each news item in turn. Each item may include content like an interview, or a report from an outside broadcast. Toward the end of the bulletin may come sports items and weather updates, finishing with a summary of the news. This gives you a sense of a well-ordered linear experience, one that an accessibility-aware engineer should strive to provide when building digital products.

Now imagine that same bulletin presented differently—we start with the weather forecast, move through news items in random order, sometimes with periods of silence, sometimes with repeated content, then hear the headlines last. The experience is illogical, incomplete, and difficult to follow, and it illustrates what happens when linearization has not worked well. Unfortunately, a screen reader experience like this is the outcome when insufficient attention has been paid to accessibility in the engineering of digital products. Users may find it difficult to understand content and may find it difficult to orient themselves within a screen or page.

In addition to audio output, linearization is also a preferred approach for some people when reading text on screen. People using screen magnification may prefer a single column of text as being easier to read and track location. Some people may prefer to remove peripheral, potentially distracting content like sidebars, navigation menus, and ads to improve focus and make reading easier. This functionality represents another form of linearization that might be offered by a reader mode in a browser or document reader, or by an eBook reading device.

A key feature of an accessible digital product is therefore that the content is logical, meaningful, and complete when linearized.

5.2.2 Keyboard-Driven Navigation and Orientation

Another usability advance that graphical user interfaces introduced was interaction with a pointer, typically controlled by a mouse or equivalent input device. A pointer allows direct manipulation of buttons, menus, and other controls, avoiding the need for users to memorize and apply keyboard commands. However, for a nonvisual screen reader user, there is limited use for an input device that relies on moving a pointer to a specific location on the display screen (or for a display screen, for that matter). Add to this the inefficiency of having to listen to all the contents of a user interface read out in sequence before being able to orientate and interact with areas of interest. Fortunately, the limitations of graphical user interfaces for nonvisual interaction can be addressed through enhanced keyboard-based interaction.

Most users are familiar to at least some extent with keyboard-based interaction with a user interface—the TAB key to move focus between interactive controls, ENTER to activate a control that currently has focus, and directional arrow keys to navigate between, for example, items in a menu or radio button list.

Screen reader users make use of those standard keyboard commands but also benefit from a wide range of additional keyboard commands provided by the screen reader that significantly improve the power and efficiency of interaction in general or with specific applications. For example, keystrokes are available to:

- Control how much content is read by the screen reader at one time, from a single character or word at a time, to line by line, to paragraphs, to the whole page in one go.
- Navigate by heading and by heading hierarchy level.
- Navigate by and within lists and tables.
- Review and navigate by interactive control type, including hyperlinks, buttons, menus, checkboxes, radio buttons, and text input fields.
- Navigate between applications currently running and between windows and tabs of the same application.
- Check the system status.
- Customize speech output, including speech rate.

In our radio news bulletin example from above, you can think of screen reader navigation keystrokes as ways that allow immediate and direct access to the start of each piece of structured content in the bulletin, for example, sports news, weather updates, or the start of the bulletin. In this way, a screen reader user can break free from the limits of linearization, for example, by skimming content structure to get a general outline of what's available and how it's laid out or quickly locating a specific piece of content of interest. Ryan Jones' account of two rooms provides a powerful alternative analogy to help understand the screen reader's navigation experience.[1]

Each screen reader has its own selection of keystrokes. If you're not a screen reader user, you don't need to learn every single one, but building a knowledge of some of the more commonly used keystrokes can help when testing for efficient screen reader navigation.

Screen readers further enhance keyboard interaction by providing a range of interaction modes, each with a distinct way in which keystrokes are interpreted as commands. These modes may be automatically set by the screen reader when it encounters specific content types in an application or may be manually changed by the user. Windows-based screen readers typically offer two distinct modes:

- Browse mode, sometimes referred to as virtual mode, is the default mode for exploring an interface. In this mode, keystrokes are intercepted by the screen reader and treated as hotkeys. For example, pressing the B key will move focus to the next button on the screen. Here, the screen reader manages interaction with the application and its content by means of an accessibility API, which we'll cover in more detail later in this section.
- Forms or focus mode, sometimes also referred to as application mode, is tailored to situations where the screen reader does not intercept keystrokes and instead the keypress is passed directly to the application to process. In this mode, if focus is on a text input field, pressing the B key would insert the 'b' character in the text input field. This means that a screen reader user can directly type content rather than having keystrokes interpreted as commands. A similar situation exists for web applications, where custom keystrokes specified by the application will be honored by the screen reader when in application mode.

VoiceOver on macOS, iOS, and iPadOS provides an additional paradigm for webpage interaction called the Rotor, a feature conceptually modeled on a dial that can be turned to different settings. Each setting is an item type appropriate to the application currently in focus. For example, in an email application, the Rotor could be set to enable quick navigation between each message in a list of emails. When a browser is open, the Rotor can be set to enable navigating from, for example, heading to heading or list to list. VoiceOver also supports two navigation modes for websites: Navigate by Document Object Model (DOM) mode supports linear interaction with content, while Navigate by Group supports interacting with specific groups of content, such as lists.

With the emergence of touchscreen devices such as smartphones and tablets, where the need for an additional physical keyboard would be inconvenient, screen reader interactions on these devices have evolved to be driven by touch gestures. When a screen reader is running on a touch-screen device, the functionality associated with gestures changes to be optimized for nonvisual interaction. This includes navigation and reading modes that operate by swiping to move linearly between items on a page or by moving a finger around the screen to hear each item announced. Operation changes as well—when a screen reader is not running, a single tap on a link or button will activate that control. When a screen reader is running and a control has focus, a double tap is needed to activate the control. This separates the actions of receiving information about the control and of activating the control.

Given the flexible and diverse ways in which screen readers provide keyboard operation for applications of different types, a key feature of an accessible digital product is therefore one that enables effective keyboard operation—both by standard keys and by screen reader keystrokes. Support for keyboard operation is also essential for other assistive technology users. As we covered in the previous chapter, *Assistive Technology*, there are a range of assistive technologies that make it easier to operate and interact with a digital product. Though each alternative input device may seem very distinct in how it looks and how it's operated, from an engineering perspective, any alternative input device that operates, for example, as a switch device will also be able to control a digital product designed for keyboard operation. Ensuring that all operable elements in the user interface can be focused and activated using a keyboard makes it possible for other input devices to operate those same elements without any extra coding effort. So a focus on keyboard access doesn't mean we should forget about mouse, touch, and speech as input methods, but accessible digital products start with effective keyboard operation.

5.3 Adaptability

One of the main attributes of an accessible product is adaptability—content and functionality that adapt to the user's context of use. We access products using different devices, for example, a web page on a laptop screen, a large monitor, or a smartphone. At one time, this adaptability of digital products was novel and innovative. Now we take it for granted.

The plasticity of digital interfaces affords greater access for people with accessibility needs than many aspects of the physical, analog world. People who are blind or have a vision impairment may not need to seek out large-print or Braille versions of books and documents when there are digital versions that are designed to be adaptable. With accessible digital documents, we can view the same document in different ways optimized to help us read, gather, and retain knowledge—using high contrast, with white text on a black background, with enlarged text—and have the document read aloud using screen reader software.

In fact, given the inherent flexibility of digital technologies, as designers and engineers, we need to think about our designs as more of a "suggestion" of how content is presented. In the same way that people may be browsing using different screen sizes that change what an optimum screen layout might be, people can choose to view pages with or without design, with large text, with different colors, and without images. Our role as digital makers is to anticipate the range of adaptations people may make and create designs that adapt gracefully to diverse user needs.

5.3.1 Separation of Content and Presentation

To support flexibility and adaptation, we need to observe the principle of separating content from how that content is presented. Visual appearance is an important characteristic of an application's interface and can contribute positively to accessibility, usability, and user experience by presenting content and functionality in recognizable ways. For example, visual appearance can communicate an organization's branding and identity and define a particular aesthetic appearance that's in line with user expectations. For many people with user accessibility needs, visual design is critical to supporting a quality user experience.

But as we learned in previous chapters on user accessibility needs and assistive technology, some people may need to adapt a visual design to meet their specific needs and make interface content easier to see, read, and understand. And some people will experience the interface nonvisually, through audio and tactile channels. Accessibility barriers can be introduced when visual characteristics are so closely integrated into the application's code that it's impossible to separate content from presentation.

One key aspect of achieving device independence and interoperability is engineering an application in a way that creates a clear distinction between content and the way the interface looks—in particular, the way content and functionality are visually represented. When a design relies on visual characteristics, meaning may be lost for anyone who doesn't experience the content the way the designer intended. Building an interface using native user interface elements that are standard for the platform, such as semantic HTML elements and native app components, helps ensure the digital product is built on a solid foundation that is ready to be styled in a way that meets both design and user needs.

Text-only and accessibility

In the early days of web accessibility efforts, design tactics like inline styles and images of text were well-meaning efforts to deal with visual design limitations of the early web. This led to content that could not be separated from presentation without a significant degree of effort, creating significant accessibility barriers. One approach website owners took was to provide text-only versions of websites as an "accessible" version of the site. This well-meaning effort avoided the difficult work of trying to tease apart sites built where content and presentation had not been separated. But the approach created a new problem: two parallel sites that both had to be maintained.

Before the advent of content management systems and automation of content updates, maintenance was manual, and frequently the "accessible" text-only site became out-of-date, with broken links and missing content. Early attempts to automate the generation of a text-only version were often unreliable. Along with the absence of video, colors, and other visual characteristics that can make content more accessible for some people, the perception quickly grew of a text-only website as a second-class alternative, which soon led to a lack of trust among people with disabilities of any separate "accessible" version of a digital product.

Given the inherent capabilities in today's digital technologies and the availability of standards and guidance for building accessible digital products, a separate "accessible" version should not be part of an accessibility strategy.

5.3.2 Encoded Semantics

For websites and applications built using markup languages, semantic markup provides a powerful way to separate content from presentation. You can use markup to identify the semantic meaning of content, whether text, media, or user interface controls. Coded semantics provide user agents like browsers and assistive technologies with a means to communicate meaning to users in a way they can perceive and understand. With the semantic structure of an interface accurately outlined using markup, you can then apply a styling language such as Cascading Style Sheets (CSS) to specify a particular visual appearance in a way that allows users to override that visual experience should they need to.

The most basic example of semantic markup is a heading. This book has chapter and section headings and subheadings that are distinguished using visual design attributes—size, weight, style, spacing. For people who read visually, these headings are signposts that guide attention through the information structure and serve as landmarks to return to. For people who read using nonvisual methods, such as a screen reader or braille device, these visual design attributes are not helpful because they only mean something when viewed in contrast with other elements. However, these relationships can be conveyed nonvisually using semantic markup to identify chapter and section headings and their hierarchy. For example, a screen reader could adjust speech output when announcing a heading or explicitly announce that text is a heading. In its simplest form, encoded semantics is the difference between marking up what something looks like and what it is. In the example that follows, we show two visually identical pieces of text, both intended to be section headings, along with the HTML code used for each heading. The first heading is marked up using a bold element (<b>), which only visually distinguishes it from other elements. The second is marked up as a level 2 heading (<h2>), which semantically distinguishes it from other elements as a heading, and also designates its level in the information structure.

This is a heading	`<b>This is a heading<b>`
This is a heading	`<h2>This is a heading</h2>`

Options for implementing semantics into code vary depending on the digital product's platform and technologies. On the web, semantics are core to Hypertext Markup Language and readily available using native HTML elements. Where HTML falls short in providing semantics, particularly for web applications, the Accessible Rich Internet Applications (WAI-ARIA) specification provides the means to add semantic attributes to native HTML elements. We'll discuss WAI-ARIA further in the context of implementing accessibility requirements.

Therefore, the task when building websites and applications is to use semantic markup fully and correctly for its intended purpose. Other coding languages represent semantic elements in different ways. For example, iOS and Android have similar but distinct ways to encode semantic properties in content and functionality, such as headings, labels, and descriptions, and in components, such as buttons, menus, and toolbars.

5.3.3 Equivalent Alternatives

A user's context of use affects what channels they have available for receiving content. For example, content provided visually is available to users who can see it, and content provided as text is available to people who can read it. A user's ability to see and read is influenced by many contextual factors, including physical and situational impairments. In these cases, adaptability means providing equivalent content in alternative formats so that people can work with the format that best suits their context.

For example, for non-text content, including images, video, and audio, perception and understanding depend on sight and hearing. People who do not have access to the required sensory channel do not have access to the content. In these cases, we must provide an equivalent alternative, with the content in another form that provides equivalent information or an equivalent experience. In most cases, an equivalent alternative will be text, but the way in which the text is provided will vary depending on the content type and purpose. And the definition of what is equivalent is also context-specific.

Depending on the platform and coding language, an image can be provided with an equivalent alternative as an attribute of the image or as an adjacent text description. For audio and video content, which is time-based rather than static, true equivalent alternatives must be synchronized with the audio or video as it plays, for example, with captions as alternatives to audio and additional spoken audio as alternatives to visual events in video.

This concept of equivalent alternatives is fundamental to building accessible digital products. In some cases, assistive technology can programmatically create and provide alternatives. For example, screen readers and reading assistance software speak text aloud as an alternative to reading. In other cases, creating and providing equivalent alternatives is part of the design and development process, for example, creating text descriptions of images and audio descriptions for video. We'll return to equivalent alternatives in more detail in Chapter 9, *Core Requirements*.

5.4 Programmatic Access to Accessibility Information

Visual user interface characteristics enable perception and understanding of content and the operation of functionality. A control may look like a button because of its shape and border and the presence of a text label or icon within the control. But this visual information is not available for assistive technologies that use alternative modalities for output, particularly the audio channel.

For nonvisual channels to be effective output and input modalities, a way is needed to communicate information about the interface, its content, and its controls in a way that can be rendered in speech and interacted with through alternative means. Critical accessibility information about a control that needs to be conveyed programmatically includes:

- **Name:** What unique function does the control perform? For example, "Save" or "Submit".
- **Role:** What kind of control is it? For example, a menu, button, or slider.
- **State:** What state is the control currently in? For example, a menu could be expanded or collapsed; a checkbox could be checked or not checked.
- **Value:** What value does the control hold? For example, a slider control could have a value anywhere between the minimum and maximum value it can hold.

Understanding what accessibility information is provided by default to an assistive technology helps you as an engineer understand:

- What code to provide in order to accurately communicate accessibility information about the control.
- What values to provide for specific attributes so that they complement what is already communicated programmatically rather than duplicate, contradict, or override valid accessibility information.

For example, in HTML, a save button labeled with an image of a floppy disk can be coded as follows:

```
<button><img src="saveicon.jpg" alt="Save"></button>
```

The button element communicates the role of the control to assistive technology, and the `alt` attribute of the image element provides the name. Together, a screen reader might announce "Button, Save". But if the `alt` attribute's value was "`Save button`", then the control's name now also includes the control's role (a button), unnecessarily duplicating information already provided programmatically. In that case, a screen reader might announce the button as "Button, Save button".

And if a control was visually labeled correctly but inadvertently provided with the wrong accessible name, then there could be severe consequences for an assistive technology user, as in the following example of a delete button with an accessible name of "Save":

```
<button><img src="deleteicon.jpg" alt="Save"></button>
```

5.4.1 Enabling Programmatic Access through the Accessibility API

So how can programmatic accessibility information about a user interface be communicated to assistive technologies? The answer is the platform accessibility API (Application Programming Interface). Since the advent of the graphical user interface presented significant accessibility challenges, platform accessibility APIs have emerged as the most robust way to represent a user interface in a way that is meaningful to an assistive technology, especially one that may be presenting and supporting interaction in a nonvisual way, such as a screen reader.[2,3]

In simple terms, accessibility APIs represent the user interface as a hierarchical structure (the "accessibility tree"), exposing accessibility information about each interface element and relationships between elements to operating systems, software, and assistive technologies. In this way, the accessibility API is used by assistive technologies to convey content and functionality to users in a meaningful way. It's what allows assistive technology to provide specialist features for navigating and operating software, such as the many hotkey commands that screen readers provide. In our example above of a control that is programmatically defined as having a role of button and an accessible name of "Save," the button can be located using a screen reader's B hotkey, which moves focus to the next button in the tree. The button can be operated with speech input because the role (button) and name ("Save") are exposed to the platform speech API. In this case, a person using speech to operate the interface could control the button through voice commands by saying, "Click Save."

Accessibility APIs are available on most modern operating systems, including desktop and mobile operating systems, and provide a robust way for accessibility information about user interface elements to be exposed to assistive technologies. For web content and applications, browsers support accessibility APIs by representing elements of a webpage as an accessibility tree, taking information from the webpage's DOM and CSS applied to the page.

It's worth noting that there may be subtle variations between platforms and assistive technologies in how effectively that information is accessed and presented. As accessibility platform support continues to evolve, assistive technologies may vary in the way they take advantage of the accessibility information exposed to them through the platform accessibility API; they may also vary in how they attempt to process incomplete or malformed accessibility information. On the web, browsers also vary in their level of accessibility support for certain HTML elements, which relates to the extent to which they expose accessibility information to the platform accessibility API and the extent to which they enable keyboard operation of an element (where the element is interactive).

This means different assistive technologies may present the same digital product in slightly different ways, and the same AT may vary in how it presents a website depending on which browser is used. This is especially true for screen readers, which, while broadly similar in how they render a user interface, have subtle differences in support. The result may be unexpected issues with the correct exposure of accessibility information, especially for custom and complex user interface elements. The community-maintained Accessibility Support website documents assistive technology support for various web technologies.[4]

As an accessibility engineer, unless you are building applications like browsers, document viewers, or assistive technologies, it's unlikely you'll need to work directly with a platform accessibility API. What's important is that you know that this mechanism exists to expose accessibility information to assistive technologies, and that for the accessibility API to do its job, the user interface needs to be coded in a way that provides the necessary accessibility information to the API.

So providing and dynamically maintaining accurate accessibility information for user interface elements is a core accessibility objective of any digital product. Using native controls wherever possible reduces the effort—and the chance of error—when exposing accessibility information to the platform accessibility API.

These core attributes comprise the foundation of any accessible digital environment. In the next chapter, *Guidelines and Standards*, we'll cover how these features and attributes are specified and supported through design principles, guidelines, and technical standards.

Every engineer should know… content and functionality must be machine readable.

By Makoto Ueki

An important keyword for making digital content more accessible is "machine readability." In this context, "machine" means the user agents such as browsers, assistive technologies like screen readers, search robots, and so on.

Take a website, as an example. A web page has several headings: a main heading that represents the subject of the web page, headings for the sections that make up the page, and subheadings within the sections. Headings are often larger in font size than the main body text and are bolded. In the case of a web page, headings can be machine readable as headings by marking them up with the HTML `<h1>` to `<h6>` elements. This allows screen reader users to browse only the headings on the page using the heading navigation function. It helps the screen readers to get an outline of the page and get to where the desired content is more quickly without having to read the page content from beginning to end.

This is a good example of accessibility that is not possible with print media such as newspapers and magazines, and can be enhanced only with digital content. Of course, the appearance of digital content is also important. And because it is digital content, it can respond to a wider range of user needs and usage environments by combining human readability with machine readability.

Let me give you another example: web pages can use images. If an image conveys some information, that information needs to be machine readable, and HTML allows the authors to set the `alt` attribute on the `<img>` element to provide alternative text that conveys the equivalent information to the user as the image. This allows users who cannot see the image on the browser screen to perceive that information by converting the alternative text described in the HTML code into audio using a screen reader or braille using a braille display.

Next, let's look at an example from a slightly different perspective. Most web pages have links and buttons; in HTML, there is a `<button>` element for a button. If the `<button>` element is used, the browser allows keyboard operation, the screen reader reads "button," and so on, to cover a variety of user environments.

However, in recent years, there have been scattered cases where buttons have been implemented in a unique way without using the standard `<button>` element, using JavaScript or other methods on the `<div>` element, for example. In many cases, these buttons are implemented so that they can be operated with a mouse, but they do not support keyboard operation or reading out loud as a button by a screen reader. A higher level of accessibility can be ensured by simply using standard HTML elements properly.

Always consider whether the page is understandable not only to the human user looking at the screen, but also to user agents such as browsers, screen readers, and even search crawlers. It will make your content more accessible to more users.

Takeaways

As an engineer, you should:

- Recognize the powerful role of device independence and interoperability in accessible digital products and seek to support these attributes.
- Pay close attention to the code order to support effective content linearization and keyboard interaction.
- Maximize platform capabilities to separate content and presentation, encode semantics, and provide equivalent alternatives.
- Know the full extent of accessibility API support that's available on platforms and technologies that you use to build digital products, and encode accessibility information so that assistive technology can render interface components.

Notes

1 R. Jones (2018) *A Tale of Two Rooms: Understanding Screen Reader Navigation.* https://www.tpgi.com/a-tale-of-two-rooms-understanding-screen-reader-navigation/
2 L. Watson and C. McCathieNevile (2015) Accessibility APIs—A Key to Web Accessibility. An Introduction to the Role of the Accessibility API. *Smashing Magazine.* www.smashingmagazine.com/2015/03/web-accessibility-with-accessibility-api
3 N. Hadder (2022) *Swinging Through the Accessibility Tree Like a Ring-Tailed Lemur—A Deep Dive into Accessibility APIs,* Part 1. knowbility.org/blog/2022/accessibility-apis-part1
4 *Accessibility Support.* a11ysupport.io

6

GUIDING PRINCIPLES

Objectives

In this chapter, we introduce established principles, guidelines, and standards that relate to digital accessibility. We describe their intended scope and objectives and the context in which they are most useful as a guide for engineering accessible technology.

Once you're through this chapter, you should:

- Understand the differences between principles, guidelines, and standards and the different use cases supported by each type of resource.
- Be familiar with the range of accessibility principles, guidelines, and standards that define and guide creation of accessible digital products.
- For each resource, understand its purpose and intent so that, as you are engineering digital products, you know which resource to use and how to apply it, based on the task at hand.

Introduction

As digital accessibility efforts have evolved over the years, collective knowledge and experience has become embedded in a range of well-established, reliable resources that encapsulate accessibility requirements and best practices. These resources will guide you as an accessibility engineer to create digital products that can be used by people with user accessibility needs. The resources fall into the following categories, based on purpose and scope:

DOI: 10.1201/9781003288060-7

- **Principles** are high-level, generic guiding statements intended to support the process of creating digital resources. They can be applied across diverse technology platforms and to diverse digital products for use in diverse situations.
- **Guidelines** provide more specific advice, which may in some cases be applicable to a specific context or technology platform.
- **Standards** are formal documents published by a standards body that may incorporate principles and guidelines. In some cases, a standard may be expressed as a regulation, where conformance with the standard is required by law.

All of the resources we introduce in this chapter are important to the practice of digital accessibility. The Web Content Accessibility Guidelines (WCAG) in particular, through its principles, guidelines, and success criteria, provide a helpful framework for exploring design and implementation requirements. We use WCAG as a structure in Chapter 9, *Core Requirements*, where we focus on WCAG requirements as ways to express accessibility requirements. In subsequent chapter, we show how WCAG can guide accessible design and development, and help teams evaluate and test whether requirements have been successfully implemented, as well as document accessibility and standard conformance in digital products.

6.1 Principles

Accessibility principles help you orient your efforts toward achieving accessibility goals. They can help teams create a shared understanding of accessibility objectives without going too deep into implementation details and guide conversations around designing solutions to meet functional requirements in an accessible way.

6.1.1 Principles of Universal Design

Some accessibility principles emerged from efforts to create accessible and inclusive physical environments and products. The Principles of Universal Design were developed in 1997 by a multidisciplinary team at North Carolina State University and remain a useful resource for digital product engineers. Universal design (UD) is defined as "the design of products and environments to be usable by all people, to the greatest extent possible, without the need for adaptation or specialized design." The seven principles of UD summarize the core characteristics of an accessible product or environment, and each has supporting guidelines.

- **Equitable Use:** The design is useful and marketable to people with diverse abilities.
- **Flexibility in Use:** The design accommodates a wide range of individual preferences and abilities.
- **Simple and Intuitive Use:** Use of the design is easy to understand, regardless of the user's experience, knowledge, language skills, or current concentration level.
- **Perceptible Information:** The design communicates necessary information effectively to the user, regardless of ambient conditions or the user's sensory abilities.
- **Tolerance for Error:** The design minimizes hazards and the adverse consequences of accidental or unintended actions.
- **Low Physical Effort:** The design can be used efficiently, and comfortably, and with a minimum of fatigue.
- **Size and Space for Approach and Use:** Appropriate size and space is provided for approach, reach, manipulation, and use regardless of user's body size, posture, or mobility.[1]

The Principles of Universal Design are known and used throughout the world, sometimes with modifications to fit the context of use and the culture and practices of the location. They can be applied to a range of contexts, including the physical environment, education, and technology, to address diverse user accessibility needs. For example, one offshoot is Universal Design for Learning, or UDL, which applies UD principles to teaching and learning.[2]

As an engineer focusing on digital accessibility, these principles provide a framework for implementing the core attributes discussed in the previous chapter. For example, the attributes of compatibility and adaptability are based on Principle 2, *Flexibility in Use*. Digital products that work with a user's assistive technology and adapt to different contexts of use support flexibility in use because the product accommodates different use cases. And because those attributes are native to the product, they also support Principle 1, *Equitable Use*, since all users can enjoy the benefits of the same product. Familiarity with the intent of each UD principle can help you avoid getting bogged down in the more technical details of accessibility. UD guides thinking and decision-making toward using integrated, design-focused solutions rather than technical workarounds to address user accessibility needs and advance digital inclusion.

6.1.2 POUR Principles

The World Wide Web Consortium (W3C) publishes the most widely recognized and adopted guidelines to support efforts to create accessible digital resources. W3C's Web Accessibility Initiative (WAI) leads efforts to define and update guidelines and supporting resources, along with efforts to ensure accessibility requirements influence other W3C specifications and resources. We'll explore the Web Content Accessibility Guidelines (WCAG) in more detail shortly, but it's useful to look separately at the underlying principles that influence W3C's accessibility efforts. These principles are referred to by the acronym POUR, which stands for Perceivable, Operable, Understandable, and Robust.

The four principles are defined as follows:

- **Perceivable:** Information and user interface components must be presentable to users in ways they can perceive.
- **Operable:** User interface components and navigation must be operable.
- **Understandable:** Information and the operation of user interface must be understandable.
- **Robust:** Content must be robust enough that it can be interpreted reliably by a wide variety of user agents, including assistive technologies. This means that users must be able to access the content as technologies advance (as technologies and user agents evolve, the content should remain accessible).[3]

The POUR principles are broadly acknowledged as the guiding principles of digital accessibility. For example, they are core to accessibility requirements for products and services covered by the European Accessibility Act (EAA), including computers and operating systems, ATMs, ticketing, check-in machines, and smartphones.[4]

For engineering digital accessibility, the POUR principles define the purpose and intent of specific accessibility requirements. They are the outcomes that result from implementing accessibility requirements—when accessibility is implemented in a digital product, it is perceivable, operable, understandable, and robust. We'll explore the principles in greater depth in Chapter 9, *Core Requirements,* where we use the POUR principles as a framework for presenting digital accessibility requirements and approaches for meeting user accessibility needs.

6.2 Guidelines

Accessibility guidelines provide more specific details on ways to implement accessibility principles and achieve accessibility outcomes. When accessibility guidelines are sufficiently detailed, they can perform two roles:

1. Help product teams design and build digital products that are accessible to people with disabilities.
2. Serve as a testing benchmark, as a resource to help ascertain whether a digital product meets a specific level of accessibility.

In the context of software engineering, guidelines provide a reliable source of accessibility requirements for digital content and functionality. They can be incorporated into specification, evaluation, and testing and used to document accessibility in digital products.

6.2.1 Universal Design Guidelines

The Principles of Universal Design we introduced earlier also provide guidelines for achieving each principle. These are relatively high-level guidelines that are adaptable to different environments, including physical and digital contexts. Here we explore a subset of guidelines that are particularly applicable in the digital context.

The guidelines under Principle 1, *Equitable Use*, aim to ensure a disabled person experiences the design in the same way as a non-disabled person, to the greatest extent possible. For digital products, following Guideline 1a, "Provide the same means of use for all users: identical whenever possible; equivalent when not," means making design decisions that favor a single accessible product rather than a separate "accessible" version or "accessibility" setting that a disabled person has to activate so they can use the product. The guideline also supports features that allow users to operate products in different ways, such as using a keyboard or speech.

Principle 2, *Flexibility in Use*, focuses on ensuring the design adapts based on user accessibility needs. Following Guideline 2a, "Provide choice in methods of use," means allowing users to adapt, for example, the visual display of a product to high-contrast mode. Following Guideline 2d, "Provide adaptability to the user's pace," means providing a mechanism to allow users to extend timeouts.

The guidelines under Principle 3, *Simple and Intuitive Use*, aim at making designs easy to understand and use, regardless of content. The guidelines encourage simple designs that make use of consistent patterns and conventions. Guideline 3e, "Provide effective prompting and feedback during and after task completion," is particularly useful in designing accessible user interactions that guide users through tasks.

In Chapter 5, *Core Attributes*, we covered the attribute of equivalent alternatives that is core to digital accessibility. Similarly, the guidelines under Principle 4, *Perceptible Information*, aim at ensuring content is presented in a format that is perceivable regardless of context or sensory abilities. This means providing content in different formats and also optimizing perceivability in providing content, for example, using high-contrast colors to support legibility.

The guidelines under Principle 5, *Tolerance for Error*, Principle 6, *Low Physical Effort*, and Principle 7, *Size and Space for Approach and Use*, apply more to physical considerations, although they can pertain to aspects of interfaces. For example, Guideline 5c, "Provide fail safe features," is a helpful guideline when designing high-risk user flows, for example, when designing transactions and allowing users to review their data before submission. The guidelines related to minimizing physical effort and providing space for use align with the core attribute of keyboard-driven navigation covered in the previous chapter, supporting keyboard users and people using alternative input devices.

These foundational concepts can help immensely when defining requirements and making design decisions, providing a design-focused, inclusive rationale and approach to creating accessible and enjoyable user experiences for everyone.

6.2.2 Web Content Accessibility Guidelines (WCAG)

Accessibility of web content is addressed through the Web Content Accessibility Guidelines (WCAG). WCAG is the most well-established and well-known of the three sets of guidelines, given that web content creators form by far the largest audience of the W3C's accessibility guidelines.

Version 1.0 of WCAG was published in 1999, reflecting the accessibility requirements of the web as it existed at that time. As the web grew in significance as a platform for information exchange, doing business, providing education, entertainment, and enabling social networking, so the nature of web technology and interactions evolved, and many of the guidelines of WCAG 1.0 became obsolete. Version 2.0 of WCAG was published in 2008 in a new structure, intended to support easier objective testing for conformance across more diverse web technologies. Additional requirements were added in version 2.1 of WCAG, published in 2018, and in version 2.2, published in 2023.

WCAG is structured in layers of increasing granularity, consisting of principles, guidelines, and success criteria. As we covered earlier, WCAG is based on four overarching principles: Perceivable, Operable, Understandable, and Robust. Each principle has one or more constituent guidelines, which provide a more general requirement for accessible web content. And each guideline has one or more constituent success criteria, with each success criterion represented as a fine-grained, testable requirement. An overview of the structure of WCAG 2.2 is outlined in Table 6.1, showing the relationship between principle and constituent guidelines and success criteria, and the topics covered by each guideline's associated success criteria.

Each success criterion is written as a statement that is either true or false, in theory allowing a tester to ascertain if the success criterion is met for a particular webpage or website. Each success criterion is also assigned a priority level based on a combination of the impact on affected users of

Table 6.1 Structure of WCAG 2.2

Principle	Guidelines	Success Criteria (SCs)
1. Perceivable	1.1. Text Alternatives	1 SC covering non-text content
	1.2. Time-based Media	9 SCs, covering alternatives for live and prerecorded video and audio
	1.3. Adaptable	6 SCs, covering topics related to programmatic communication of content structure and relationships, purpose of inputs and interface controls, and reading order, along with use of sensory characteristics in instructions, and display orientation management
	1.4. Distinguishable	13 SCs, covering use of color, color contrast, use of images of text, audio control, flexibility of text when properties are changed, and management of content that shows on hover or focus
2. Operable	2.1. Keyboard accessible	4 SCs, covering keyboard operation of functionality
	2.2. Enough time	6 SCs, covering control over moving content and timeouts
	2.3. Seizures and physical reactions	3 SCs, covering flashing and animated content
	2.4. Navigable	13 SCs, covering keyboard focus display, appearance, and management, identification of link purpose, use of descriptive headings and labels, and supporting flexibility in navigation
	2.5. Input modalities	8 SCs, covering pointer behavior, size of controls, use of motion actuation and dragging movements for functionality, and identification of labels and names for controls

(Continued)

Table 6.1 (*Continued*) Structure of WCAG 2.2

Principle	Guidelines	Success Criteria (SCs)
3. Understandable	3.1. Readable	6 SCs, covering programmatic identification of content language, reading level, treatment of unusual words and abbreviations
	3.2. Predictable	6 SCs, covering consistency in identification, location, and behavior of interactive controls and help mechanisms
	3.3. Input assistance	9 SCs, covering error identification and recovery support, quality of labels and instructions, and accessible authentication and data entry
4. Robust	4.1. Compatible	2 SCs, covering programmatic identification of name, role, state, and value information for interactive elements, and identification of notifications

About W3C accessibility guidelines

W3C accessibility guidelines are the most widely known and commonly used of all digital accessibility guidelines. W3C's approach to providing guidelines for web accessibility is based on a three-part model of responsibility for accessibility:

1. **Web Content:** Authors are responsible for following best practices in creating accessible web content.
2. **Authoring Tools:** The tools that authors use to create web content must support authors of all abilities in creating accessible web content.
3. **User Agents:** The applications used by people to access and interact with web content must ensure that people with disabilities can access and interact with web content.

W3C has published accessibility guidelines for each part of the model, and these guidelines work together to collectively set requirements that need to be met to enable accessible interactions with web content and applications.

In W3C terminology, guidelines are published as a "recommendation," and each W3C recommendation is referred to as a "normative" document—a standard in all but name and the result of a formal process of authoring and review. W3C publishes many supporting "informative" documents, which explain how to apply normative requirements.

not meeting the criterion and the technical effort required to meet the criterion. These priority levels are expressed as Level A, Level AA, and Level AAA. Since publication, around the world, and at the time of writing, WCAG 2 (version 2.0 or 2.1) Level AA has become the *de facto* conformance level for digital accessibility.

W3C publishes a number of supporting resources for WCAG, aimed at all audiences expected to use WCAG. Some of the most helpful resources include:

- "How to Meet WCAG" is a customizable reference that shows all associated resources with each constituent SC and allows filtering of WCAG's requirements based on a range of categories.[5]
- "Understanding WCAG" is a collection of documents supporting each SC. These include detailed explanations of the SC's purpose in terms of user groups who benefit and the specific user accessibility needs that are met. The resources also include examples of techniques with which the SC may be satisfied, as well as techniques that, if implemented, would lead to failure to meet the SC's requirements.[6]

As an engineer, knowledge of WCAG is a significant part of understanding and applying digital accessibility best practices on the web and beyond. We cover designing, developing, and testing WCAG requirements in Part 2, *Methods for Engineering Digital Accessibility*.

6.2.3 Authoring Tool Accessibility Guidelines

The Authoring Tool Accessibility Guidelines (ATAG) provide requirements for any tool whose purpose is to generate or support the generation of web content. Authoring tool features allow users to generate, create, or edit content, such as creating webpages using a website building tool or exporting content as a PDF document. In many cases, authoring tools support authors in creating content and functionality without requiring much, if any, understanding of code or programming logic. Some examples include:

- A blogging tool that enables authors to create or edit blog posts using a template for specifying post title, metadata, content, and styling.
- A learning management system with functionality that allows instructors to design tests that ask questions using text, images, and video content and request students to provide answers through either free text responses, uploading video or audio files, or selecting from predefined answer options.
- A system for human resources staff to manage staff records of employees and applicants for open positions, which can generate custom PDF reports of data stored in the system.

- A cloud-based application that allows users to design their own website based on a set of predefined templates without needing to write any code.
- A social media application that enables the sharing of text messages, images, video, and audio with selected recipients or openly to anyone.
- Software for creating content with text, images, video, and audio, such as an email or word processing application.
- Design and development tools, such as tools for creating software prototypes and development environments used to engineer web applications.

Advances in authoring tools have helped make it easier for people to create and share web content. However, the availability of authoring tools has also increased the proportion of web authors who are creating web content without knowledge or skills in accessibility.

Historically, web authoring tools lacked the support to allow authors to easily generate accessible content. In some cases, functionality to support accessible authoring existed but was provided in an obscure or advanced setting, which meant few authors would be aware of or know how to use this functionality. In response, W3C published ATAG as a set of functional requirements for authoring tools in an attempt to ease the process of creating accessible content for all users, regardless of prior knowledge. Version 2.0 of ATAG, published in 2015,[7] follows a similar structure to WCAG, presenting principles, guidelines, and success criteria.

ATAG is presented in two parts, with complementary objectives:

1. Part A focuses on guidelines for making the authoring tool user interface accessible to all authors, regardless of disability. Part A incorporates the requirements of WCAG by reference and aims to ensure people with user accessibility needs can produce web content.
2. Part B focuses on supporting the production of accessible content by all authors, regardless of accessibility knowledge and skills. This section includes requirements for functionality that support authors applying accessibility best practices in authoring decisions, easing the process of generating accessible code, preserving accessibility features in content import, export, and management activities, and performing accessibility checks.

As an engineer, you should be familiar with the requirements of ATAG, especially if you are involved in building digital products that support content authoring. Ensuring people who have user accessibility needs can use authoring tools as content authors as well as use the content produced by authoring tools is integral to digital inclusion for disabled people.

6.2.4 User Agent Accessibility Guidelines

The third set of guidelines is perhaps the least well-known of the W3C's accessibility guidelines and almost certainly the least observed. The W3C defines "user agent" as "any software that retrieves, renders, and facilitates end-user interaction with web content."[8] This includes applications that are platform-based, embedded user agents such as plugins, and web-based software such as browsers. User agents include desktop and mobile browsers, media players, document viewers, functionality that enables native mobile apps to render web content, and authoring tool functionality that can preview authored web content. In short, a user agent is a means for web content to be exposed to a web user, including someone with accessibility needs. If the user agent does not provide a means for a web user to access accessibility features of the web content, then it may not be possible for that user to experience the content in an accessible way.

The User Agent Accessibility Guidelines (UAAG) provide functional requirements for user agents to ensure that they have the functionality needed to present web content in a way that exposes accessibility information and features to users, with the aid of assistive technology where needed. UAAG version 2.0 was published as a W3C recommendation in 2015[9] and follows a similar structure to WCAG, presenting principles, guidelines, and success criteria.

UAAG's five constituent principles are:

1. Ensure that the user interface and rendered content are perceivable.
2. Ensure that the user interface is operable.
3. Ensure that the user interface is understandable.
4. Facilitate programmatic access.
5. Comply with applicable specifications and conventions.

As an engineer, you should be familiar with the requirements of UAAG, especially if you are involved in building a browser, media player, document viewer, or any digital product that enables the presentation of web content.

6.2.5 Other W3C Resources

Although WCAG was originally created as a resource to support web content accessibility, its success has led to efforts to apply WCAG to other digital platforms and resource types. While WCAG 1.0 was written to be applicable to HTML-based websites, version 2 of WCAG involved significant efforts to make the guidelines more platform-agnostic. Even with this effort, applying WCAG to non-web platforms can be challenging, given that some concepts introduced in WCAG, such as user agents, content, and web pages, may not be applicable to other platforms and technologies. The range in availability and capability of hardware, operating systems, and assistive technologies and how each works together to enable access to and interaction with a digital resource can make it difficult to interpret how best to apply a specific WCAG requirement.

Additionally, WCAG was originally aimed at "content," and the user experience of the web has extended from consuming content to producing and interacting with content in complex and dynamic ways. Equally, today's technology stack is far more complex and multi-dimensional, as needed to provide users with the interactive experience we've all come to expect. Also, early versions of WCAG largely focused on requirements for nonvisual web access, but there are many additional user accessibility needs that digital accessibility guidelines and standards must address.

To fill these gaps, there are various other W3C accessibility efforts aimed at extending WCAG, ATAG, and UAAG to support accessibility in a broader range of digital products and services for a more diverse range of user accessibility needs. Here, we cover a few of those efforts.

WAI-ARIA is a W3C Recommendation that specifies supplementary HTML attributes and attribute values for the purpose of providing information about controls, interactions, and dynamic content to assistive technology. WAI-ARIA, which is shorthand for the "Accessible Rich Internet Applications Suite," was first published in 2014 to fill a gap in semantic support for assistive technology with the advent of dynamic content and interactions and complex, custom components and widgets. WAI-ARIA allows web authors to programmatically define roles and states for user interface components that aren't native HTML elements. For example, an HTML <select> element is standard HTML, but HTML does not have a native dropdown element like those commonly used for site navigation menus. Adding accessibility semantics through ARIA attributes like role and aria-expanded to HTML elements allows authors to provide essential details about user interface (UI) components to assistive technology. When these attributes are correctly implemented, tools like screen reader software can convey the necessary information to navigate and operate interactive elements. The accompanying ARIA Authoring Practices Guide (APG)[10] provides detailed implementation support for common UI patterns, including accordions, menus, and tabs. If you build HTML, you should be adept at using ARIA to enhance UI accessibility and should consult the APG when building custom interactive components.

WCAG2ICT is a W3C Working Group Note providing guidance on applying WCAG to non-web technologies and resources, including desktop software and native mobile applications. Published in 2013 by the WCAG2ICT Task Force, WCAG2ICT, short for Guidance on Applying WCAG 2.0 to Non-Web Information and Communications Technologies, the guidance is intended for digital product management, design and development, testing and evaluation, and policy-making.[11] This resource takes the intent of each WCAG 2.0 success criterion and provides an explanation of how the intent of each SC may be met on non-web platforms. At the time of writing in 2023, the WCAG2ICT Task Force is working to update the document to cover more recent versions of WCAG. Familiarity with WCAG2ICT and the work of the WCAG2ICT Task Force helps you extend WCAG's requirements to other technology platforms.

Content Usable is another helpful W3C Working Group Note, developed by the Cognitive and Learning Disabilities Accessibility Task Force (COGA) to provide greater support for cognitive user accessibility needs. Short for "Making Content Usable for People with Cognitive and Learning Disabilities," Content Usable provides personas and scenarios, user stories, and design patterns to help designers and engineers understand what's needed to improve digital accessibility for people with cognitive and learning disabilities. The objectives in Content Usable come from another W3C resource, Cognitive Accessibility Guidance. The guidance contains objectives, such as "Help Users Avoid Mistakes and Know How to Correct Them," and associated design patterns, such as "Let Users Go Back." The guidance and techniques in these resources are a rich source of insight for designing and developing user interfaces that meet a broad range of cognitive user accessibility needs.

6.3 Standards

Accessibility standards are published by an established standards body, detailing requirements for implementing accessibility in digital products. They provide measures that can be used to determine whether digital products meet a consistent baseline for accessibility. Standards are often used as part of policy and regulatory contexts and applied across groups and organizations to hold them accountable for maintaining a common baseline.

With web accessibility, the most commonly referenced standard is the Web Content Accessibility Guidelines, version 2.0 of which is also an international standard known as ISO/IEC 40500:2012, *Information technology—W3C Web Content Accessibility Guidelines (WCAG) 2.0*. Other accessibility standards include the *ICT Accessibility 508 Standards and 255 Guidelines in the United States* (Section 508) and the *Accessibility Requirements for Products and Services in the European Union* (EN 301 549). In the United States, in 2023, work is underway to provide the Americans with Disabilities Act with a standard for ADA compliance in the digital environment, complementing a standard for accessibility in the physical world first published in 1991. These standards incorporate WCAG principles, guidelines, and success criteria in one form or another.

Organizational policy may reference specific standards for web accessibility and indicate conformance levels for policy compliance. For example, your company might have an accessibility policy that states something like, "Our digital products and services conform to Web Content Accessibility Guidelines version 2.2, Level AA (WCAG 2.2 Level AA)." As an employee of the company, you will be expected to comply with organizational policy, which in this case means designing and developing digital products that conform to WCAG 2.2 Level AA.

Some standards go beyond web accessibility to include requirements for hardware, software, and documents. Both the ICT Accessibility Standards in the U.S. and EN 301 549 in the EU include requirements for hardware and software, communication technology, and documentation and support services. The ADA Standards for Accessible Design include requirements for closed systems, such as self-service transaction machines and kiosks. ISO 14289, *PDF Enhancement for Accessibility* (also referred to as PDF/UA) has requirements for producing accessible PDF documents.

Some accessibility standards focus on standardizing the *process* of building digital products rather than the output—the product being created. Following a process standard increases the chances that the development process will lead to the outcome of a more accessible product. One such process standard is ISO/IEC 30071-1:2019, *Information technology—Development of user interface accessibility—Part 1: Code of practice for creating accessible ICT products and services.*

In summary, there are many accessibility standards defining how you work and what you create. As an engineer, you should know of their existence, be ready to meet accessibility requirements in the digital products that you design, implement, and manage, and encourage their adoption by your product team and your organization. One of the best ways to ensure you are prepared to meet accessibility standards is by understanding your accessibility role and adopting accessibility processes and practices. In the next chapter, we explore what it means to be an accessibility-aware engineer with an accessibility-informed professional practice.

Takeaways

As an engineer, you should:

- Be conversant in accessible design principles and be ready to use them to guide accessibility discussions and decision-making toward intentional, designed outcomes.
- Know and apply the requirements specified in the Web Content Accessibility Guidelines to digital content and interactions of all kinds, including non-web interfaces.

- Draw on other relevant guidelines and standards that are best suited to your context, such as ATAG for working on digital products that allow users to author content.
- Commit to staying up to date with developments in accessibility guidelines and standards that apply to your professional focus area.
- Consider joining working groups or other initiatives to develop accessibility standards as a way of staying current and involved in advancing disability inclusion.
- Learn the accessibility expectations relevant to your location, your team, and your organization.

Notes

1 *Principles of Universal Design* (1997). Compiled by advocates of Universal Design in 1997. Participants are listed in alphabetical order: Bettye Rose Connell, Mike Jones, Ron Mace, Jim Mueller, Abir Mullick, Elaine Ostroff, Jon Sanford, Ed Steinfeld, Molly Story, Gregg Vanderheiden. The Principles are copyrighted to the Center for Universal Design, School of Design, State University of North Carolina at Raleigh, USA. universaldesign.ie/about-universal-design/the-7-principles
2 CAST (2018). *Universal Design for Learning Guidelines version 2.2.* udlguidelines.cast.org
3 World Wide Web Consortium (W3C) (2023). *Web Content Accessibility Guidelines.* www.w3.org/TR/WCAG
4 *Directive (EU) 2019/882 of the European Parliament and of the Council of 17 April 2019 on the Accessibility Requirements for Products and Services.* data.europa.eu/eli/dir/2019/882/oj
5 *How to Meet WCAG (Quick Reference).* www.w3.org/WAI/WCAG21/quickref
6 *WCAG 2.1 Understanding Docs.* www.w3.org/WAI/WCAG21/Understanding
7 *Authoring Tool Accessibility Guidelines (ATAG) 2.0.* www.w3.org/TR/ATAG20
8 Definition of User Agent, from *Appendix A: Glossary of User Agent Accessibility Guidelines (UAAG) 2.0.* www.w3.org/TR/UAAG20/#def-user-agent
9 *User Agent Accessibility Guidelines (UAAG).* www.w3.org/TR/UAAG
10 *ARIA Authoring Practices Guide (APG).* www.w3.org/WAI/ARIA/apg
11 *Guidance on Applying WCAG 2.0 to Non-Web Information and Communications Technologies.* www.w3.org/TR/wcag2ict

7

ACCESSIBILITY IN PRACTICE

Objectives

In this chapter, we explore digital accessibility roles and responsibilities and examine what might be needed on a personal, team, and organizational level to fully include accessibility in your professional practice. We discuss personal and contextual challenges that may hinder accessibility in practice and ways to advance accessibility through personal and professional development and organizational advancement.

Once you're through this chapter, you should:

- Understand your role as an accessibility-aware engineer; recognize barriers that may limit success in fulfilling your role and ways to overcome them.
- Be familiar with accessibility roles and responsibilities for members of the product team.
- Recognize a range of approaches and options for building competency and capacity.
- Understand how to utilize processes and procedures to advance the maturity of your own practice, for your team, and in your organization.

Introduction

As you embark on your journey toward becoming a more accessibility-aware engineer, it's worth preparing yourself for the many challenges you'll encounter along the way so that you have the skills and strategies to address those challenges—whether they are presented by organizations or other people, or they exist within yourself.

DOI: 10.1201/9781003288060-8

We hope that by reading this book, you will gain accessibility knowledge and skills that will help you optimize the accessibility of the digital products you're involved in creating. Over time, we hope that accessibility becomes sufficiently embedded in processes and practices that it becomes a standard expected quality of all digital products. Most organizations are not at that point yet, which means you're highly likely to encounter situations where existing processes are inadequate to ensure optimal accessibility.

As you build your knowledge and skills in designing and developing accessible digital products, you'll likely encounter the reality that there are challenges to achieving accessibility goals at your organization. Your organization might be more reactive than proactive in its approach to accessibility. Maybe the project sponsor doesn't really understand that accessibility covers more than visual impairments. Maybe the product manager doesn't see accessibility as a priority unless it's specifically called out in the product requirements. Maybe the development framework chosen doesn't make it easy to provide user interface elements with the appropriate accessibility information. Maybe accessibility suddenly became mandated when a potential customer demanded it, and your team is now having to figure out what level of accessibility they can include without causing the project timeline or budget to slip.

This means that your accessibility skills and knowledge will likely be applied in a transformation scenario, one where you'll be working with suboptimal processes and tools and where accessibility may not be seen as the highest priority. This is the reality we all need to face, and it's our intention as authors that this book will help equip you with the skills to help your organization transform into one with greater accessibility maturity. But before you can be an effective agent of change, you need to maximize your readiness to adopt a practice of accessibility.

7.1 Your Accessibility Role

Accessibility brings both responsibilities and opportunities to engineers who are building digital products. What does it mean to be an accessibility-aware engineer? In our opinion, the answer is a combination of strategic and practical knowledge and skills, a philosophy that embraces diversity, and a commitment to overcoming barriers to achieving accessibility goals.

7.1.1 Foundational Knowledge and Skills

You need a broad understanding of the range of user accessibility needs that exist and the hardware and software that people with disabilities have at their disposal to enable the use of digital technology. You also need an appreciation of the range of support for accessibility that exists across different technology platforms and how that impacts what you build and how. You also need a foundation in the concepts of digital accessibility and the principles, guidelines, and standards that distill what we already know to be robust ways to avoid introducing accessibility barriers, meet accessibility requirements, and create an inclusive user experience.

Building on these core concepts, you need an understanding of how responsibility for accessibility is best distributed across different phases of the digital product lifecycle. This includes how accessibility influences requirement establishment, design, development, and content creation. It includes ways in which accessibility can be effectively incorporated into testing and evaluation activities throughout the development lifecycle and how accessibility is effectively managed before and after launch.

You need an understanding of how to efficiently and effectively integrate accessibility into research, design, development, and testing efforts, and you need tooling to support those efforts. You also need an awareness of your organization's current capacity for accessibility and of effective ways to manage situations when others fall short in meeting their accessibility responsibilities.

Overarching all of the practical skills and knowledge is a commitment to focusing on understanding and designing for user diversity and the recognition that accessibility is an ongoing effort of continuous improvement rather than a one-time effort.

7.1.2 An Inclusive Mindset

If you don't have lived experience of disability, as a disabled person or through close relationships with disabled people, you may have an incomplete or inaccurate understanding of what it's like to live with a disability. This may influence your attitudes and biases about people's accessibility needs and how to design for them. And if you identify as disabled, you may be all too familiar with these attitudes and biases and the barriers they produce; you may also recognize that your lived experience isn't representative of all other disabled people.

- Some people may see disability as a group of medical conditions that are accommodated or fixed through support and rehabilitation services rather than as an integral part of human existence that affects everyone.
- Some people's understanding of accessibility may be grounded in the physical world, and they may be unaware of the role of digital technologies in overcoming impairments. They may think accessibility is about curb cuts and access ramps, not websites and apps.

- Some people may have knowledge gaps about accessibility and disability inclusion that cause them to question whether it's worth the effort. Is digital accessibility really a problem for people? How many people are likely to be affected?
- Some people may accept accessibility as part of their digital role but see it as a burden, requiring extra time and effort and limiting creativity and opportunities to innovate.

In addition to assumptions about disability, technology, and accessibility, you may have implicit and explicit biases about disability that affect how you approach accessibility. You may think disabled people have impairments that limit their ability to use technology, reducing your motivation to address accessibility in your work because you aren't familiar with the range of assistive technologies and accessibility strategies. You may find it uncomfortable to think about disability, making it a struggle to take the time to understand and connect with the experience of disabled people.

Even within the disability world, there are biases that can influence attitudes, as we alluded to when we introduced user accessibility needs in Chapter 3. Mark Deal's theory of the hierarchy of impairments describes the phenomenon whereby some disabilities warrant more attention and focus than others, influenced by perceptions of empathy and sympathy.[1] In this theory, people may be more likely to want to help address issues related to blindness and vision impairment, compared to deafness, hard of hearing, or cognitive impairments. This might help explain why, in the digital world, there is a tendency to equate accessibility with making access possible for screen reader users, and subsequently, discussion and resources that focus on accessibility tend to be dominated by screen reader accessibility, with less (or no) attention paid to other accessibility needs.

Assuming responsibility for minimizing inaccessibility for all disability groups through careful design will improve your engineering practice and provide the context for innovation and excellence. But like everything else, prioritizing accessibility comes with trade-offs. For example, building an interface that is self-explanatory, with input labels and instructions that are visible, may look cluttered to some people. Budgeting time for captioning media may be difficult when weighed against pressing release dates and priorities. Most people would agree with the position that disabled people have the same rights as nondisabled people, including access to the digital world. However, when it comes to taking action to support equity by prioritizing accessibility over other factors in product development, you may find barriers that get in the way.

Addressing attitudinal barriers requires that you change your attitudes and address your biases. One way to do that is by identifying motivators for change and acting on them. There are positive motivators since accessibility and competency benefit your professional profile.

- **Accessibility is part of an ethical practice.** A major motivator for ensuring digital products are usable by people with disabilities is the idea that technological advances should help solve problems that exist in the world, including social exclusion. As a technology professional, do you want to be part of the problem or part of the solution? Do you want to cause harm, or do you want to open doors to opportunity and participation?
- **Accessibility is a requirement for usability.** Engineering programs historically have tended to emphasize technical qualities such as performance and security over human factors such as usability and user experience. Understanding how design decisions could affect potential users, including people who have accessibility needs, will help you be a better engineer.
- **Accessibility is a professional skill.** With accessibility competency, you will be more ready to assume new roles (or meet the changing requirements of an existing role) as your career evolves. An engineer applicant who is versed in accessibility will be more hireable and ready to perform the responsibilities of the role.

When technology creates new barriers for people who already disproportionately encounter barriers and exclusion, we exacerbate rather than improve the situation. Thus, understanding the risks of a lack of accessibility can also be a powerful motivator. As a technology professional, you can cause direct and indirect harm by not paying attention to accessibility needs and requirements.

- **Lack of accessibility can cause direct harm to disabled people.** For example, some technologies cause physical harm, such as seizures and migraines caused by flashing content, and pain and fatigue due to interaction. Some technologies cause emotional harm, such as anxiety due to unexpected changes to content or a lack of instructions to complete a critical task.
- **Lack of accessibility can indirectly cause harm to disabled people.** Some technologies cause various forms of indirect harm by preventing access to, for example, critical services, employment, and education due to design and engineering oversights that result in barriers.

A philosophy of accessibility awareness means paying regular attention to the potential impact of design and engineering decisions on accessibility and encouraging others to share in that attention. As you become more accessibility-aware in your practice, you may find the need to engage constructively both with laggards and resisters as well as advocates who may feel that not enough is being done. We encourage you to keep learning, keep encouraging others, and look for opportunities to raise awareness, identify process improvements, share knowledge and skills, and celebrate achievements, no matter how small they may be.

Every engineer should know... those who have been marginalized get it.

By Jonee Meiser

All too often, I run into development teams that continuously treat accessibility solely as technical bugs that they will address whenever they get to them. The impact of those existing bugs is perceived as minimal. And that is when I have been known to lose my patience with teams. But when I lose my patience, I encounter the stereotypes that then exclude me from the conversation—and exclude me from participation. What is ironic about that, though, is how folks do not even recognize the irony.

My passion for accessibility is often perceived as me being the "Angry Black Woman" or me having a bad attitude. I am not a team player because I am speaking against the product. Imagine that. I face tone-policing during my fight for someone else's inclusion. I get mansplained what "Beta" and what "Minimum Viable Product" means and how testing an inaccessible product or service for market fit does not exclude people with disabilities. Gaslighting is real. Taking the time to justify why accessibility will not be addressed at any given time is called privilege, my friend. Wonder why I am passionate about accessibility? Because I get it. As a Black woman, I get what it feels like to be excluded from society and to be denied access to places and spaces where I know I belong. I have been told my natural hair is inappropriate. Heck, I have been told my experiences altogether as a Black woman are inappropriate to discuss at work. I am a doctoral candidate and have been working since I was a senior in college. Yet, I see people in positions of power with way less professional experience and education, which telegraphs to me that I am not welcome, that people like me are not welcome. People who do not fit the description of the majority are not welcome. This is what inaccessibility telegraphs to people with disabilities: that they are not welcome in your digital environment.

Here is what I want you to know. I simply want you to know that inclusion and belonging are at the heart of web and digital accessibility. When we leave accessibility bugs in our digital and web-based products, services, and the like, we exclude people with disabilities from the conversation and participation. When we leave accessibility bugs where they are, we are denying people with disabilities access to digital places and spaces where they have a right to belong. We are not talking minor technical bugs here for the developers to fix. We are talking discrimination against people with disabilities. We are talking about the social responsibility of ensuring that folks have fair and equal access. It is just something I want you to think about the next time I lose my patience with you when you tell me you will address it after you release it in production.

7.2 Team Roles and Practices

There was a time where one person could manage multiple facets of product development—design, coding, maintenance—especially with websites and apps. Today's technology development is a social, collaborative practice. Most technology professionals work on a multi-disciplinary product team, with colleagues in overlapping roles and responsibilities collaborating on tasks and learning from one another. It's certainly more complex and chaotic than a person designing and developing their own vision and specification, which characterized the early World Wide Web. But as technology becomes even more critical to people and societies, we all benefit from the diverse and multiple viewpoints of creators and users of these systems that we rely on for our daily lives.

Similarly, there has been a significant shift in how we develop and sustain technological products. There was also a time when the development process was more of a production line, with each person working on an assigned task within their area of expertise, which they would then hand off to the next person. Today's development process is more integrated, flowing through stages and tasks, focusing on continuous improvement, and each person's daily work varies depending on the task at hand. This change to more fluid development processes supports the fluid nature of technology products, supporting creativity and innovation while also maintaining stability in core systems.

7.2.1 Shared Responsibility for Accessibility

There's been a trend over the years to see accessibility as overwhelmingly the responsibility of development. As the people writing the code that will run the digital resource, it's true that developers have a significant responsibility to ensure the code they write meets expected accessibility standards. But the reality is that many decisions affecting accessibility are made in software development long before developers start coding. A conscientious developer who wants to write the most accessible application may have limited ability to reverse the negative accessibility impact of decisions made earlier in the product lifecycle, even for projects that follow an Agile approach. Is it any wonder that accessibility can be perceived as something technologically complex and expensive to do, if there's an assumption that it's the responsibility of developers alone?

Another common approach is to delegate accessibility to an accessibility specialist or outsource accessibility to an accessibility consultancy. This approach positions accessibility as a delegated task for accessibility specialists, with issues managed as bugs to be remediated. While this approach can yield positive results in the short term, it puts accessibility responsibility on a single point of failure, making it vulnerable to cost-cutting decisions that lead to the departure of the accessibility specialist. This approach does not adequately address accessibility as a non-functional requirement that must be considered throughout product development.

The most effective approach to including accessibility as a quality attribute and non-functional requirement in digital product creation is to treat it as a shared responsibility. This may require a deliberate, staged approach to building accessibility capacity across the team, bringing everyone to a point where they understand and can meet their responsibilities.

Here we provide a range of roles with details about their accessibility responsibilities. This section considers various roles on the product development team and describes their accessibility responsibilities and accountabilities within their scope of influence and authority.

- **Product Sponsors** have the budgetary authority to fund accessible product development. They are responsible for establishing accessibility as a priority and funding accessibility, both in time and budget allocation. The project sponsor is accountable for the accessibility of the product and may need to defend accessibility against other potentially conflicting priorities.
- **Product Owners/Managers** have the authority to set accessibility priorities for product development and define development processes. They are responsible for ensuring accessibility is prioritized and managed as a non-functional requirement and compliance-level concern, establishing development processes and resources that support quality attention to accessibility. The product owner is accountable for how accessibility is prioritized and supported in product development.
- **User Experience (UX) Researchers** have the authority to include perspectives and experiences of disabled people in research data, whether to provide insights to inform new product development or to evaluate existing products or products in development. They are responsible for recruiting participants with disabilities, adapting research methods as needed to ensure equal participation, and reporting research-based accessibility insights and considerations. They are accountable for accurately capturing disability and accessibility perspectives that lead to improved accessibility, as well as ethical and inclusive research practices.
- **Business Analysts** have the authority to surface accessibility requirements for consideration in product development based on research and business goals. They are responsible for ensuring accessibility is prioritized among functional and technical requirements and correctly considered when defining project requirements. They are accountable for ensuring the needs of disabled people are addressed in project requirements.
- **Project Managers** have the authority to manage accessibility in product development activities. They are responsible for ensuring that team members understand their accessibility responsibilities and have what they need to meet their responsibilities, including training and tools. They are responsible for ensuring that accessibility

requirements are treated with equal importance as other requirements, especially when conflicts arise, and are accountable for making sure team members have the necessary skills and schedule to implement accessibility.

- **Software Architects** have the authority to make high-level decisions about technologies, tools, and methodologies that are most appropriate to building a product that meets specified requirements within budget and on schedule. They are responsible for ensuring that these high-level decisions are influenced by a project's accessibility goals, and choices are made that smooth the path toward meeting accessibility requirements rather than introducing new barriers.
- **Content Producers** have the authority to include accessibility when designing, creating, producing, and publishing content. They are responsible for ensuring content is created and published following accessible practices, including the use of clear language and semantic markup, and providing accessible alternatives for audio, video, and images. They are accountable for the accessibility of the content.
- **Designers** have the authority to include accessibility features and functionality in visual and interaction designs. They are responsible for specifying accessibility requirements in design specifications, mock-ups, and other design artifacts. They are accountable for accessible design.
- **Developers** have the authority to include accessibility when developing sites and applications. They are responsible for ensuring content and functionality are implemented to meet accessibility requirements. They are accountable for accessible features and functionality.
- **Testers** have the authority to include accessibility tests in quality assurance activities. They are responsible for evaluating whether products meet accessibility requirements and reporting the results. They are accountable for the accurate reporting of the accessibility status of products.

There are other key accessibility roles that are not directly related to product development but have a significant impact on accessibility outcomes, including:

- **Marketing and Communications** are responsible for producing accessible marketing material promoting the product, including to people with disabilities, and for highlighting the product's accessibility features in appropriate language.
- **Technical Writers** are responsible for ensuring that product documentation highlights accessibility features, identifies alternative ways to activate functionality and receive information, promotes accessibility best practices in using the digital product, and is written in language that does not marginalize users with disabilities.

- **Customer Support** is responsible for sensitively handling accessibility-related issues, including having appropriate processes for gathering sufficient customer information to identify and either address or escalate an accessibility-related issue. This may involve tiered levels of support, providing progressively more specialist accessibility support.
- **Human Resources** is responsible for recruiting employees with the requisite skills to contribute to a successful project, including accessibility skills. Human Resources is also responsible for supporting employees with disabilities by providing an accessible work environment.

Every engineer should know… accessibility is a team effort. It is not exclusive to user research or front-end development.

By Yasmine Elglaly

Accessibility is still often viewed as a niche area in software development, where only a select few individuals—such as user researchers and front-end developers—are expected to have knowledge and expertise in this field. These individuals may be tasked with not only identifying and addressing accessibility issues but also with educating other team members about accessibility, advocating for accessibility, and fostering an inclusive culture. Additionally, they may encounter resistance from team members who are skeptical about the importance of accessibility, which can make their job even more challenging. However, this approach is problematic for two key reasons. Firstly, this places a heavy burden on those with accessibility experience to build a minimum level of accessibility foundation into their teams, which takes away from their ability to focus on technical tasks such as running usability studies with users with disabilities. Secondly, accessibility is a team effort and cannot be the sole responsibility of a few individuals. It is not practical to ask one person to handle accessibility for the entire software development process, from design to implementation to testing. Every software engineer, regardless of their role, should be aware of what current tools and technologies have to offer in terms of accessibility support, as well as what is still lacking in this area.

The decisions made during the early stages of software development can greatly impact the accessibility of the software. It is crucial for software architects to have a strong understanding of the accessibility support offered by various frameworks and tools, so that they can make informed decisions on how to create accessible software. Additionally, recognizing the limitations of current frameworks and tools can help guide appropriate actions at various stages of the software development process to address potential accessibility issues.

Product owners should understand that accessibility is an essential factor that should be considered throughout the product creation process. Therefore, every user story should be checked for accessibility before it is deemed done. When accessibility is considered at the beginning of the project, the timeline will be more realistic, avoiding the unfair outcome that making software accessible is "too much work" or will take "too much time." Neglecting accessibility during project planning often leads to finding accessibility bugs during the testing phase, and making changes at this point can be time-consuming and expensive for production.

Quality assurance engineers should realize that complying with accessibility laws is not enough to ensure high-quality software for people with disabilities. Focusing solely on compliance often results in software that is challenging for users with disabilities to use. Accessibility is not just about meeting legal requirements; it is about creating software that is easy to use, enjoyable, and inclusive for all users. To ensure that software is accessible to users with disabilities, it is important to test the software with various types of assistive technology and to gather feedback from users with disabilities. By testing software with assistive technology, developers can ensure that the software can be used with screen readers, magnifiers, and other assistive devices. Additionally, testing with users with disabilities provides valuable feedback on the usability and accessibility of the software, allowing developers to make necessary improvements to ensure an inclusive user experience.

As such, it is important to build a culture of accessibility within software development teams, where everyone is encouraged and empowered to play a role in making software more accessible to all users. This requires education and training for all team members, as well as ongoing communication and collaboration throughout the development process. By involving everyone in the process of creating accessible software, teams can create more inclusive and equitable products that serve the needs of all users.

A shared responsibility for accessibility is the ideal philosophy for a product team, one where everyone understands their role and has the knowledge, skills, and tools to meet their responsibilities. In reality, it may be difficult to fully share responsibility across a team, especially when access to knowledge, skills, and enthusiasm is unevenly distributed. This might be the case in a team that collectively is new to accessibility or where the organizational culture is to see accessibility as a lower priority than other goals, leading to practices that are not optimized for achieving accessibility outcomes.

7.2.2 Proactive Practices

Like other quality attributes, including reliability, security, and safety, accessibility has multiple dimensions and layers, and must be infused into all areas of product development. It's not a checklist task that can be delegated to a single accessibility specialist on the product team at a specified development phase. Rather, it's a quality of a system that must be considered by every member of the product team at every stage of design and development, on most every task.

Unfortunately, often accessibility isn't given the same status and priority as other quality attributes. Instead, you may encounter one of the following scenarios:

- **Accessibility is Absent.** Look in your software engineering textbook and check the index. Is there an entry for accessibility or disability? How many pages are devoted to accessibility? At the time we're writing, in 2023, awareness and prioritization of digital accessibility are certainly growing, but are still lacking in many contexts related to technology product development. Depending on when you are reading this book and where you are in the world, it may not be an integral part of your technology education, a workplace priority, on the radar of leadership or project sponsors, or a consideration in product development.
- **Accessibility is Reactive.** Many products have been built following development processes that do not explicitly include accessibility considerations, where accessibility features that are present are included only by chance, and defects are only surfaced very late in the development lifecycle or after launch, when someone asks for evidence of accessibility. This might be when a stakeholder or customer asks a sales representative about a product's level of accessibility, when a user encounters accessibility barriers, or when an attorney files a lawsuit alleging disability discrimination.
- **Accessibility Comes Too Late.** Some teams include accessibility in development processes but wait to consider requirements until QA and testing. The design and development teams do not consider accessibility initially. Then, during testing, accessibility issues surface and must be managed as defects. In this scenario, reworking designs to address bugs is not typically the first recourse. Therefore, developers are often tasked with addressing accessibility defects behind the scenes, in the code.

In scenarios like these, your organization might hire an accessibility consultant to evaluate the product against accessibility standards to find out whether any issues are present. If you haven't considered accessibility in product development, it's almost certain that the product won't conform to accessibility standards. Imagine building a town library without accessibility guidelines and only consulting specifications for accessibility after the building is open and in use. Would you expect to have exactly met the clearances for wheelchair maneuvering? Would you have instinctively known which door fixtures would work for people who have limited reach and strength? Not taking accessibility into consideration means that these elements likely exclude people on the basis of disability and therefore have to be redone or worked around to meet client and customer requirements.

Often, the issues are so embedded in the foundation of the project that only incremental changes are possible. Accessibility improvements are deferred until the next system redesign, and in the meantime, people are harmed (for example, cannot use the product to apply for work), and customers are lost (for example, cannot purchase products).

When accessibility is treated as a reactive activity, the result is often "accessibility fails." These are instances where accessibility features are added to an existing design in an unsuccessful attempt to overcome barriers or address accessibility requirements after the fact. You might have experienced or encountered classic fails in the physical world, such as too-steep ramps built over stairs or obstacles like trees or planters on accessible routes.

In the digital world, accessibility failures can take a number of forms. One of the most egregious failures is to avoid remediating a product and instead rely on the provision of an alternative version of a product for people who have accessibility needs. Not only does this fail to address the underlying process issues that led to the inaccessible product, but the alternative version is highly likely to provide an unsatisfactory experience for disabled people.

Don't let these discouraging facts and cautionary tales put you off. If you take a proactive approach to accessibility, you will find that focusing on accessibility as a core value and quality attribute improves your work and your outputs overall. Constraints encourage creativity. Accessibility-first development processes allow for elegant designs with integrated accessibility features, and accessibility becomes a marker of quality to be celebrated (rather than an awkward "fail" or exclusionary barrier). Adopting a proactive approach to accessibility in your own practice raises the profile of accessibility overall. Addressing accessibility in your own work and celebrating accessibility wins helps build awareness and competency in others.

7.3 Building Awareness and Competency

As an engineer, you may find you need to work with team members who have less accessibility awareness than you might like. There's a fine balance between helping others grow their knowledge while delivering quality code that meets requirements within project constraints. Sometimes, you may have to focus on incremental improvement, encouraging team members to take slow, deliberate steps to introduce a new aspect of accessibility in to their work and building success over time.

It's true that when people aren't meeting job or role expectations, they need to be informed so that they can take action to improve. Where possible, encouraging progress and celebrating achievements is far more likely to create a positive atmosphere, a more motivated team, a shared appreciation of accessibility as a quality that everyone can contribute to, and a more rewarding outcome for accessibility advocates.

An accessibility-awareness approach brings together a commitment to people-focused design with a commitment to robust code that accommodates diverse interaction methods and minimizes reliance on untested assumptions about users and user needs. Accessibility awareness rejects the dichotomy of "normal users" and "disabled users" and recognizes that we all exist on a multi-dimensional continuum of capability. It also recognizes that we don't know everything there is to know about our users, and encourages each of us to look outside of ourselves, be curious, and learn from others.

7.3.1 Professional Development

Accessibility knowledge and skills development can take place in many different ways, and taking a hybrid approach helps meet different attitudes, roles, and learning styles. For example, there are many online accessibility trainings in self-guided or instructor-led format, covering different aspects of accessibility, from introductions to highly role- or topic-specific material. There's a wide range of YouTube and Vimeo videos on accessibility, including demonstrations of how disabled people use technology and short explanatory videos of specific accessibility concepts applied in practice. Online learning platforms provide general and specific courses on accessibility. Around the world and virtually, there are meetups, bootcamps, hackathons, and virtual and in-person conferences that are either dedicated to accessibility or include accessibility topics.

You might work in an organization where in-person or virtual lunch-and-learn or after-work events are regular occurrences, and these can be great opportunities to discuss an accessibility-related topic while also building a community of accessibility practitioners who can learn from each other. You could use these events to bring in a guest speaker to talk about a specific topic or invite disabled technology users to explain how they use assistive technology when using the web, software, and mobile apps. The importance of a real-world perspective on accessibility helps balance the abstract nature of accessibility laws, standards, and guidelines. Someone who finds it difficult to engage with the purpose or requirements of a specific accessibility guideline may find clarity when a person with accessibility needs explains why products designed to meet that requirement work so well for them.

Some people may see accessibility as an abstract concept that interferes with their preferred approach to design or coding. One way to make accessibility more real is to create opportunities for development teams to observe people with disabilities interact with products they've built—initially to see the impact of accessibility barriers that are present, and, ideally, later on to see the impact of removing those barriers. This concept of "exposure hours" as a measure of the amount of time a product team observes user research activities was coined by UX specialist Jared Spool as a valuable way to connect product teams with the reality of how users interact with their products.[2] When a developer observes someone struggle to use a product they helped create, there is an increased understanding of the context and impact of an issue, which helps build ownership of the issue and motivation to fix the problem.

7.3.2 Collaboration and Communication

Within the product team, team members should have a shared understanding of accessibility—what it is, who it's for, and why it matters—and use a common vocabulary when discussing accessibility standards and features. It's easy for people to have their own definitions of what "accessible" means, which can create misunderstandings and gaps between the current state and the desired state. Expressing accessibility goals and requirements in plain language with appropriate reference to standards and guidelines helps ensure the team is working from a common understanding.

Determine which practices work best for your team. For example, in some cases, it might be best to engage with only the accessibility requirements that are needed at the time, based on the functionality the team is focusing on. In that case, it might be overwhelming to get detailed feedback on accessibility related to features that aren't currently being developed. In other instances, teams might want to know and log all the accessibility issues and then manage them through, for example, Jira tickets.

Work to establish an effective division of responsibility so that each role is able to engage fully in the way that they find most effective. For example, when producing design annotations, some annotations might be for development to decide rather than for design to annotate and development to execute. Include accessibility in retrospectives to evaluate and refine your approach. For example, for design annotations, discuss how annotations worked and didn't work and decide on ways to adapt the process. The key is to adapt and refine the approach that works best for the team.[3]

7.3.3 Accessibility Community

You are not alone in your journey toward being a more accessibility-aware engineer. As accessibility grows from a niche profession into a topic recognized as something that any digital professional should have a level of mastery over, so too does the number of resources, organizations, and groups supporting skills development and knowledge exchange. This includes a growing number of ways in which you can connect with people with accessibility needs and partner in efforts to engineer accessible products.

In a mature and inclusive organization that recognizes accessibility as key to equality and non-discrimination, you will have a diverse network of colleagues in different roles who are paying attention to accessibility and working to fulfill their accessibility roles. Some organizations have an "accessibility champions" network of people who stay up-to-date on current standards and best practices, sharing their knowledge with each other, and bringing the knowledge and skills back to their teams. These peer networks help to establish an accessibility community of practice and are an excellent way to connect with others working to solve accessibility problems, to learn and build skills and knowledge, as well as support others in understanding and fulfilling their accessibility roles.

Looking outside the organization, there is a growing community of accessibility advocates and specialists who write and talk about ways to advance accessibility, from design and development techniques to using tools in innovative ways to improving processes and practices in a way that enhances accessibility. If you find yourself as a lone voice advocating for accessibility, you're not in fact alone! There's plenty of support beyond your organization.

7.3.4 Disability Representation

As an engineer, you may have limited influence on recruitment practices at your organization. But recognize that the more people with disabilities are part of your organization, the more apparent user accessibility needs will become. And, with more disabled employees, the richer the source of personal perspectives and feedback on product design decisions.

Your organization may have programs to increase the number of disabled employees—this is an opportunity to talk to program leaders about the value of disabled engineers, product managers, and other roles responsible for building digital products. If you have the opportunity to influence hiring within your team, emphasize the value that a disabled co-worker could bring to accessibility efforts.

7.4 Accessibility Program Management

An important tool in the accessibility engineer's toolkit is the ability to estimate the accessibility maturity of an organization—assess the limitations of projects, processes, and tools, and identify how best to make positive steps with accessibility within these limitations. With deliberate attention and a collaborative approach, you can build partnerships with co-workers and find ways to inject accessibility into flawed processes. And in doing so, you may be able to positively influence these processes so that the changes you make are lasting.[4]

7.4.1 Investing in Accessibility

Many organizations are still in the early stages of including accessibility in their processes and practices, which can make it difficult to prioritize accessibility, especially against competing priorities. Your employer or customers may be unaware that digital accessibility is a consideration or not recognize that disabled people use their products and services. The organization may believe there are no issues because there haven't been accessibility complaints. As a result, accessibility might not be a priority for your employer or manager or a requirement for your team. This means that accessibility requirements might not be included in project requirements, and you may not be allocated time for accessibility due diligence.

As a technology professional advocating for accessibility in these contexts, you may find your arguments sympathetically heard but ultimately lacking in persuasion as the organization prioritizes other goals. Trying to promote accessibility as a quality attribute when it isn't valued by your employer, manager, or team can be a challenging situation, making it difficult to successfully integrate accessibility into the planning, development, and testing of digital products.

A recurring challenge for anyone wanting to advance accessibility is the idea that building an accessible digital product adds to project costs. This is a contentious argument, often based on past experience of cost and effort to address accessibility at a late stage in a project, and one that can be used to argue that accessibility is "too expensive." It's true that some activities that deliver accessibility require additional time and effort. Producing an accessible video requires budgeting for quality captioning and, in many cases, audio description. Not including captioning or audio description would save costs and time, at the expense of accessibility for people who are deaf or hard of hearing and people who are blind or visually impaired, and at the expense of others who would benefit from these features. We may have saved time and money, but at what other cost?

As we've previously discussed, when we leave accessibility to the end of the development lifecycle, where an accessibility audit is conducted shortly before launch, the chances are there will be significant barriers present in the product. Fixing those barriers is likely to require complex redevelopment work that will take time, add to project costs, and threaten delivery schedules. In such a situation, it's easy to see how accessibility gains a reputation of a costly addition to project budgets.

The unfortunate reality is that there are costs associated with processes that fail to adequately include accessibility in digital products. These costs may relate to essential remediation work or legal costs associated with defending claims of unlawful discrimination against people with disabilities. An organization may make another choice—to invest in improving processes and practices to ensure that accessibility is adequately addressed throughout the design and development process. Either scenario can lead to accessibility's unwarranted reputation as a significant financial burden.

7.4.2 Embedding Accessibility

It's helpful to look at the physical world, where similar challenges exist in creating accessible built environments. As accessibility expectations have grown over time, driven by civil rights movements toward inclusion and equality and associated legal and regulatory requirements, so there has been an increased focus on how we can effectively improve the accessibility of the built environment in a manageable, affordable way. In the US, the ADA standards take a sensible approach to distinguishing two key scenarios: existing buildings that lack full accessibility and new building projects. Building accessibility into the start of a new building project is a requirement. Expectations for accessibility should be high when a building exists as an idea. But retrofitting an existing building is a very different challenge, and full accessibility may not be achievable.

So we can also think in similar ways about the digital world, both in terms of processes and products. If you're involved in a brand new organization with brand new processes and practices, then there's no excuse not to embed accessibility into every stage of the product development lifecycle. If you join an organization that already has processes and practices in place and already has digital products, then the likelihood is that these processes and practices will need to change to accommodate accessibility requirements more effectively and these products will have to change to meet accessibility obligations.

7.4.3 Risk Management

Accessibility barriers in digital products affect different groups and dimensions and must be managed, first by identifying who or what might be affected by the barriers and then by planning for how to manage and mitigate risks.

Risks to People: Users of products with accessibility defects may experience harm by encountering barriers that prevent them from completing tasks. Given our reliance on technology for essential activities, these barriers could cause harm by, for example, preventing access to health services. Barriers may cause harm to safety and privacy, for example, when a user is unable to independently access personal information due to accessibility defects. Users may experience exclusion, blocked by barriers that prevent them from participating in employment, education, transportation, health services, and more.

Customers of products with accessibility defects may be exposed to discriminatory practices. Disabled people are protected by civil rights laws in many locations, and access barriers constitute disability discrimination. Customers that use inaccessible products to provide programs and services may inadvertently set up barriers for disabled people. For example, a university that purchases an inaccessible learning management system to deliver education programs may discriminate against disabled people who cannot use the system by excluding them from participating in education programs.

Risks to Product and Project: Projects that must work around accessibility defects and shortcomings experience several risks, including risks to resource and schedule management, requirements, staffing, and morale. When accessibility defects exist in a product, teams may have difficulty staying on target with the project schedule and budget. Remediating defects may cost time and require staffing increments or engagement of accessibility specialists, which takes time and funding. It may be difficult to find the necessary expertise to address defects. Technical debt from accessibility bugs and defects can cost time, money, and effort, and put projects at risk. Staffing issues may arise, for example, if accessibility is delegated to one team member who then leaves, potentially due to a lack of engagement from the rest of the team.

Business Risks: Products brought to market with accessibility defects will not meet the accessibility requirements of some customers. For example, in the U.S., federal government procurement policy requires that government agencies seek out and purchase products that conform with the ICT Accessibility 508 Standards and 255 Guidelines: "Compliance with these standards is mandatory for Federal agencies subject to Section 508 of the Rehabilitation Act of 1973, as amended (29 U.S.C. 794d)."[5] Similar requirements exist for organizations in the European Union. In these cases, the business may lose customers due to incompatibility with customer procurement policies. If competing products do comply with accessible procurement requirements, that could affect market share or even force the business to take new directions. Businesses may be implicated in lawsuits related to disability discrimination or harm caused by inaccessible products. Lawsuits have financial and reputational risks.

For project management, be mindful of risks that might affect the timeline and resource management, including staffing. Ensure you have staff with accessibility skills on the project team. Don't delegate responsibility for accessibility to a single specialist, without the authority to ensure accessibility is delivered. Check on accessibility throughout the project to ensure requirements continue to be considered and addressed by the development team. If anything arises that may impact accessibility, communicate the risks to the product owner and management, along with plans to minimize the effect of carrying the risk on the project.

As an engineer, focus on risk management in product design and development, making sure to include accessibility issues in requirements engineering. Minimize the impact of accessibility defects by incorporating accessibility requirements and building to accessibility standards. Communicate any risks in implementing accessibility to project management.

For product owners, make sure accessibility standards are met and accessibility features are prioritized in product development. Communicate with customers about accessibility features and address any defects through product updates. Don't ignore accessibility issues in favor of other feature updates any more than you would let security lapse or open doors to privacy violations. Be aware of defects that exist and plan for how to address and monitor risk moving forward. For product team leads, provide staff training to maintain accessibility capacity. Make sure team members know how much value business and product management places on accessibility. Celebrate accessibility achievements and motivate team members to excel at accessibility.

7.4.4 Change Management

One of the most important tasks of software development project management is to anticipate and manage the impact of changes imposed on the project. Change could be a reduction in budget, reallocation of staff, reduced time available to build and test the product, or late changes in requirements for the product being built. All of these changes can threaten the chances of successfully delivering a product that meets accessibility requirements.

For example, a reduction in budget or time may lead to increased pressure to reduce efforts to test for accessibility barriers or remediate known barriers. A reallocation of staff may lead to loss of accessibility expertise in a team if knowledge and skills are limited to one or two team members. A change in team management or sponsorship may bring in someone who is more skeptical about the value of accessibility, and who may be less enthusiastic about approving or prioritizing efforts spent on accessibility. A change in requirements may mean a request for functionality that will be challenging to implement in an accessible way, with the staff and time available.

By development lifecycle, we refer to a set of key activities—requirements establishment, design, development, and testing. Each activity lends itself to accessibility integration, regardless of how it might be performed. An even distribution of accessibility throughout the lifecycle helps reduce the burden of development and testing, reduce the cost of fixing accessibility issues discovered later in the lifecycle, and increase the chances of an end product that meets its accessibility requirement.

In Part 2 of this book, *Methods for Engineering Digital Accessibility*, we focus on accessibility in the development lifecycle, from requirements through to maintenance. We provide details about what every accessibility-aware engineer should know and address in each of these project phases, so that accessibility is a proactive, people- and quality-focused effort that produces accessible digital products.

Takeaways

As an engineer, you should:

- Make accessibility a professional priority and commit to establishing and sustaining a foundation of accessibility knowledge and skills.
- Commit to continuous work on overcoming attitudinal and operational barriers, within yourself and others, so that they do not get in the way of advancing your accessibility practice.
- Recognize your accessibility role and the roles of your colleagues and team members, and share accessibility responsibilities accordingly.
- Approach accessibility as a proactive, quality-focused activity and address accessibility concerns at optimal stages in your work so that your efforts have the greatest impact on accessible outcomes.
- Engage in accessibility with your colleagues, organization, and professional communities; be an active participant in accessibility communities of practice to learn and grow along with others.
- Be persistent in your accessibility attention and advocacy, despite any barriers and challenges that you may encounter.

Notes

1 M. Deal (2003) Disabled people's attitudes toward other impairment groups: A hierarchy of impairments. *Disability & Society*, vol. 18, no. 7, pp. 897–910. https://www.tandfonline.com/doi/abs/10.1080/0968759032000127317
2 J. Spool (2011) *Fast Path to a Great UX—Increased Exposure Hours*. articles. centercentre.com/user_exposure_hours
3 D. Barrell (2020) *Agile Accessibility Handbook: A Practical Guide to Accessible Software Development at Scale*. Herndon, VA: Amplify Publishing.
4 S. Byrne-Haber (2021) Giving a damn about accessibility. *UX Collective*. www.accessibility.uxdesign.cc
5 *ICT Accessibility 508 Standards and 255 Guidelines*. www.access-board.gov/ict

Part 2
Methods for Engineering
Digital Accessibility

The task of addressing accessibility in digital products begins at the beginning—in planning, discovery, and specification. All too often, accessibility is an afterthought and is approached as a technical compliance activity of testing, bug tracking, and remediation. It's no wonder that accessibility can be perceived as difficult and limiting in technology development. In Part 2, *Methods for Engineering Digital Accessibility,* we present a proactive, holistic approach to accessibility, with creative, user-centered methods and tasks to be performed by all roles on the product team in all phases of the product lifecycle.

- Chapter 8, *Requirements Specification*, presents approaches for surfacing and defining accessibility requirements and ways to describe them meaningfully in requirements documentation.
- Chapter 9, *Core Requirements*, uses global accessibility principles and guidelines as a framework for describing core requirements, their purpose and intent, and considerations for meeting them in design and implementation.
- Chapter 10, *Design and Development*, provides approaches to designing and implementing accessibility requirements.
- Chapter 11, *Testing and Evaluation*, covers automated and manual accessibility testing and accessibility evaluation methods, as well as ways for recording and acting on issues that are identified.
- Chapter 12, *Documentation and Support*, focuses on best practices for documenting the status of accessibility in digital products and considerations when communicating accessibility status and supporting end users.

DOI: 10.1201/9781003288060-9

With a solid grounding in digital accessibility foundations and methods, we move to the final chapter, Chapter 13, *The Future of Digital Accessibility*, and invite you to step forward as an accessibility-informed and competent design and technology professional, ready to contribute to engineering an accessible and inclusive digital world.

8

REQUIREMENTS SPECIFICATION

Objectives

In this chapter, we present guidance on how to incorporate accessibility requirements into project specifications. We discuss how accessibility can be treated as a non-functional requirement (a system constraint or quality) and as a functional requirement. We explore ways to express accessibility requirements to best suit a particular product and development approach.

Once you're through this chapter, you should know:

- The difference between accessibility requirements as functional requirements and non-functional requirements.
- Ways to articulate accessibility requirements that are relevant to the product being developed.

Introduction

In digital product development, requirements elicitation and specification are critical aspects, guiding design and implementation activities in directions that best meet business goals and user needs. Without this phase, product viability is at risk. When accessibility is not included in a product's requirements, there's an increased probability that the product will not be usable by people with accessibility needs and will not meet business requirements for accessibility. This makes it important to ensure that, regardless of how or when product requirements are specified, accessibility is included in the requirements specification. Accessibility requirements describe specific content and functionality that a product must provide to be usable by people with accessibility needs. They also describe constraints the product must work within, such as those imposed by standards, laws, and business imperatives.

DOI: 10.1201/9781003288060-10

This chapter explores different approaches and activities related to defining and articulating accessibility requirements. We focus on two primary activities—requirements discovery and requirements specification—and cover considerations and approaches. But first, we explore common paths toward defining accessibility requirements through conformance with standards and exploration and articulation of accessibility needs through user research. We also examine how accessibility fits in with other functional and non-functional requirements.

8.1 Approaches to Accessibility Requirements

Implementing requirements is more straightforward when requirements are measurable and testable, making it possible to validate that they have been satisfied in design and implementation. This is one reason accessibility is often expressed as a requirement that a product conforms to an accessibility standard, for example, "The product will conform with WCAG 2.2 Level A and AA standards." In this way, the requirement can be met and validated when evidence of conformance is available, for example, through test results or a conformance audit.

When defining accessibility requirements, we can look to standards such as the Web Content Accessibility Guidelines (WCAG) for a baseline technical specification of accessibility requirements that we can use as a foundation for all systems, and thus include conformance with WCAG as a *de facto* requirement. The standard provides the foundation and detailed specifications for implementing accessibility in digital products. Its primary scope is web technology, but the underlying intent of WCAG's requirements can be extrapolated to apply in general to digital interfaces. WCAG is also a common measure used throughout the world to assess whether a digital product has barriers that may cause disability discrimination. In many organizations, WCAG conformance is a business requirement, specified in the laws and regulations of the country or region or sector, organizational policies and procedures, and business agreements and contracts with clients, customers, and consumers.

Standards are, by nature, generic expressions of accessibility requirements. They may not provide sufficient detail to ensure that the specific product being built meets an optimum level of accessibility and usability for disabled users. When we approach accessibility solely as a standards-conformance exercise, we miss out on many of the benefits and opportunities that come with focusing on making products work for people with accessibility needs. So product requirements should also focus on what's needed to ensure that functionality is provided in a way that provides value to disabled people and provides an optimal user experience. This requires careful consideration of the product context—who is using the product, for what purpose, where, and when—along with user research into critical needs in terms of what functionality is provided and how.

Some product quality attributes are core to accessibility, including flexibility, adaptability, device independence, encoded semantics, and native controls, as covered in Chapter 5, *Core Attributes*. At the requirements stage, we can prioritize these attributes and translate them into product-specific requirements to carry forward into design and implementation, as well as avoid adding conflicting or unnecessary requirements. Attention to user accessibility needs in the requirements phase and careful documentation of the requirements will reduce the need for extensive testing and remediation of accessibility defects later in the development lifecycle, and reduce technical debt.

Every engineer should know... not all user needs or barriers will be addressed by the "current" WCAG guidelines at any given time.

By Erich Manser

One thing I consider essential for engineers who are concerned with digital accessibility to understand is that the WCAG (Web Content Accessibility Guidelines) Standards, while foundational and extremely important, are also quite limited in scope and coverage.

Even as WCAG continues to evolve and expand, there will always be use cases in which some will face digital barriers that simply are not covered by standards, especially as new technologies continue to emerge. Standards provide a technical basis and clear guidance, but we must also take continual guidance from the shared experiences and real needs of people with disabilities.

For example, I am someone with low vision who relies on magnification and inverted colors in order to see what's on my screen. The appearance of light text on a dark background is softer and less glaring for me, and enhances my ability to perceive what's there, and view it without pain or discomfort.

Recently, however, I've noticed more interfaces that combine both light and dark themes in a single user experience, giving a split light-dark appearance, almost like the traditional Chinese Yin-Yang symbol. From my perspective as a user, this is problematic because it automatically means there is some portion of that page or app that I will be unable to view comfortably, yet there is nothing in current WCAG guidelines to require color uniformity across a user interface.

While this is a single, personal experience, I know many others with disabilities who have similar experiences with barriers not covered by WCAG.

8.2 Types of Requirements

In his book, *Software Engineering*, Ian Sommerville defines functional and non-functional requirements as follows:

> Functional requirements (…) are statements of services the system should provide, how the system should react to particular inputs, and how the system should behave in particular situations. In some cases, the functional requirements may also explicitly state what the system should not do.
>
> Non-functional requirements (…) are constraints on the services or functions offered by the system. They include timing constraints, constraints on the development process, and constraints imposed by standards. Non-functional requirements often apply to the system as a whole rather than individual system features or services.[1]

Non-functional requirements are also referred to as "quality attributes" that are used to evaluate the performance and operation of a system. Like other quality attributes, such as security, reliability, and performance, many accessibility requirements don't relate specifically to a type of function or service provided to users but rather to how the system is provided. With some non-functional requirements, meeting requirements can be a matter of complying with technical standards. As a non-functional requirement, accessibility is a quality of the digital product system-wide, and meeting accessibility requirements means the website, kiosk, app, software, or other digital product is accessible because it conforms with technical accessibility standards.

In addition, systems have functional requirements related to accessibility. For example, if you are developing a system that includes image upload functionality, you need to design the system so that the content has alternative descriptions that meet your non-functional requirements for accessibility, which include requirements that provide nonvisual access to visual information. Your user requirements may be, "Users can add alternative text descriptions when uploading images." This user requirement has significant system implications that include providing functionality for users to add a description, perhaps requiring that users add a description when uploading an image, or maybe auto-generating image descriptions and then allowing for editing, storing image descriptions, allowing updates to image descriptions, and so on.

When it comes to accessibility requirements, we need to approach accessibility as both a non-functional requirement and functional requirement. We need to embed accessibility into user requirements (what services the system will provide to users) and system requirements (what services and functions will be implemented). And we need to specify requirements that support accessibility and avoid other requirements that may conflict with accessibility requirements.

Dealing with conflicting requirements

The process of specifying requirements often leads to discovery of conflicts between different sets of requirements. This can happen with accessibility requirements. Often, the conflict exists due to implementation choices rather than a core underlying conflict. In other words, the conflict is due to how the requirements were prioritized and acted upon in design and implementation. For example, a requirement to display text at a fixed size and layout is in conflict with adaptability as covered in Chapter 5, *Core Attributes,* allowing layouts to adapt to accommodate different text sizes based on user settings.

A good example is the apparent conflict between accessibility and security requirements that require some way to authenticate a user as a human as part of a login process, rather than a piece of software trying to access a system for potentially malicious purposes. This may be a legitimate business security requirement, to reduce risk of data breaches or ransomware attack.

A common solution is the CAPTCHA feature, which typically sets a user a test, in theory one that a human should find straightforward but which a script may not. Unfortunately, some implementations of CAPTCHA use tests that assume humans have certain visual, intellectual, reading, processing, problem-solving or motor capabilities (such as the ability to visually identify specific images from a selection). The consequence of such tests is that not only are non-humans filtered out, but also some humans with disabilities. Thus, the security requirement may have been met, but implemented in a way that does not meet accessibility requirements.

An accessibility-aware product team might consider alternative ways to meet the security requirement while ensuring that people with disabilities can still access and use the system. The W3C publication Inaccessibility of CAPTCHA[2] provides some suggestions for accessible approaches to secure authentication processes.

In summary, treat accessibility requirements with equivalent priority to other requirements, and where conflicts appear to arise, conduct careful exploration of how alternative approaches can be taken to satisfy the requirements that appear to be in conflict.

8.3 Requirements Discovery

Defining accessibility requirements starts with discovery, approaching the task from different directions—legal requirements, user requirements, business requirements, and more—to arrive at an understanding of what's needed and viable pathways toward meeting those needs. Successful accessibility outcomes are largely predicated on the extent and effectiveness of the discovery process. All too often, meeting accessibility requirements is approached as an after-the-fact exercise of remediating issues based on standards conformance audits. Prioritizing accessibility in discovery and articulating those requirements for effective design and implementation is key to digital products that are not only accessible to but also of value to people with disabilities.

8.3.1 Business Requirements for Accessibility

Input for a digital product's user requirements for accessibility may come from different sources. One source is the business requirements, the expression of what a business needs the product to do so that it will deliver value to the business. Value might be defined as sales figures if the product is being built to sell to customers. If the product is being built for internal use, value might be defined as, for example, productivity improvements or cost savings. Business requirements can inform functional requirements in terms of features that should be offered to users for the product to provide the value expected by the business.

When business requirements include accessibility, these requirements may capture the business's recognition that accessibility is a desirable quality of the product to be built. In our experience, it's more likely that business requirements for accessibility reflect the business's obligations to comply with local or national disability rights legislation. Some organizations may have an internal accessibility policy that is referenced in business requirements of any digital product the organization builds. The policy may require conformance with specific accessibility standards for all digital product development and may include operational process requirements relating to technology procurement or development.

Sometimes business requirements are influenced by legal responsibilities of customers of the product being built. For example, if the product's intended customers include federal agencies or federally funded organizations in the United States, customers will need evidence that the product conforms to the Section 508 Standards. But in other cases, you may find accessibility is absent from business requirements, particularly where the stakeholders sponsoring or influencing the direction of the product are not aware of or do not value accessibility.

Standards for other aspects of accessibility also have established specifications that can be included in business accessibility requirements. For example, businesses may require that product development processes follow specific standards to maximize efficiency, manage quality, and mitigate risks. As we noted earlier, the standard ISO/IEC 30071, *Code of practice for creating accessible ICT products and services*, includes guidance on defining and following a process for building and maintaining accessible digital products.

8.3.2 User Requirements for Accessibility

Complementing the business influence on requirements should be the influence of the needs of users expected to use the system. For accessibility, some standards can help us with a basic articulation of user accessibility

Table 8.1 Functional Performance Criteria in Section 508 and EN 301549

Section 508	EN 301549
302.1 Without Vision. Where a visual mode of operation is provided, ICT shall provide at least one mode of operation that does not require user vision.	4.2.1 Usage without vision. Where ICT provides visual modes of operation, the ICT provides at least one mode of operation that does not require vision. This is essential for users without vision and benefits many more users in different situations.
302.2 With Limited Vision. Where a visual mode of operation is provided, ICT shall provide at least one mode of operation that enables users to make use of limited vision.	4.2.2 Usage with limited vision. Where ICT provides visual modes of operation, the ICT provides features that enable users to make better use of their limited vision. This is essential for users with limited vision and benefits many more users in different situations.
302.3 Without Perception of Color. Where a visual mode of operation is provided, ICT shall provide at least one visual mode of operation that does not require user perception of color.	4.2.3 Usage without perception of color. Where ICT provides visual modes of operation, the ICT provides a visual mode of operation that does not require user perception of color. This is essential for users with limited color perception and benefits many more users in different situations.
302.4 Without Hearing. Where an audible mode of operation is provided, ICT shall provide at least one mode of operation that does not require user hearing.	4.2.4 Usage without hearing. Where ICT provides auditory modes of operation, the ICT provides at least one mode of operation that does not require hearing. This is essential for users without hearing and benefits many more users in different situations.
302.5 With Limited Hearing. Where an audible mode of operation is provided, ICT shall provide at least one mode of operation that enables users to make use of limited hearing.	4.2.5 Usage with limited hearing. Where ICT provides auditory modes of operation, the ICT provides enhanced audio features. This is essential for users with limited hearing and benefits many more users in different situations.
302.6 Without Speech. Where speech is used for input, control, or operation, ICT shall provide at least one mode of operation that does not require user speech.	4.2.6 Usage with no or limited vocal capability. Where ICT requires vocal input from users, the ICT provides at least one mode of operation that does not require them to generate vocal output. This is essential users with no or limited vocal capability and benefits many more users in different situations.

(Continued)

Table 8.1 (*Continued*) Functional Performance Criteria in Section 508 and EN 301549

Section 508	EN 301549
302.7 With Limited Manipulation. Where a manual mode of operation is provided, ICT shall provide at least one mode of operation that does not require fine motor control or simultaneous manual operations.	4.2.7 Usage with limited manipulation or strength. Where ICT requires manual actions, the ICT provides features that enable users to make use of the ICT through alternative actions not requiring manipulation, simultaneous action or hand strength. This is essential for users with limited manipulation or strength and benefits many more users in different situations.
302.8 With Limited Reach and Strength. Where a manual mode of operation is provided, ICT shall provide at least one mode of operation that is operable with limited reach and limited strength.	4.2.8 Usage with limited reach. Where ICT products are free-standing or installed, all the elements required for operation will need to be within reach of all users. This is essential for users with limited reach and benefits many more users in different situations.
No equivalent.	4.2.9 Minimize photosensitive seizure triggers. Where ICT provides visual modes of operation, the ICT provides at least one mode of operation that minimizes the potential for triggering photosensitive seizures. This is essential for users with photosensitive seizure triggers.
302.9 With Limited Language, Cognitive, and Learning Abilities. ICT shall provide features making its use by individuals with limited cognitive, language, and learning abilities simpler and easier.	4.2.10 Usage with limited cognition, language or learning. The ICT provides features and/or presentation that makes it simpler and easier to understand, operate and use. This is essential for users with limited cognition, language or learning, and benefits many more users in different situations.
No equivalent.	4.2.11 Privacy. Where ICT provides features for accessibility, the ICT maintains the privacy of users of these features at the same level as other users.

needs through functional performance criteria. Section 508 and EN 301549 both include a similar set of functional performance criteria, listed in Table 8.1. You can think of these criteria as baseline requirements that need to be met for disabled people to be able to use a product, regardless of the product's type, purpose, or context of use.

Functional performance criteria are useful as a baseline for accessibility requirements, covering a broad spectrum of user accessibility needs. But because they are generic, they are insufficient on their own to provide the level of detailed requirements needed to guide building a product for a specific purpose using a specific technology platform. For this level of detail, we need more input. An important source of functional requirements is up-front research into user needs—in particular, with people who are current or potential users of the product being built. Insights gained from learning first-hand about functional needs and solutions from people who have expertise and lived experience in these areas are invaluable at the requirements stage. This is the moment to capture and incorporate requirements that address user accessibility needs, help the product team consider creative ways to meet those needs, and ultimately guide development in the right direction.

User research focuses on gathering rich data from current and potential users about behaviors and needs in order to inform the design of a product that is useful to the target audience. This type of research complements market research, which, in simple terms, focuses on gathering data around what customers want. User research can yield valuable insights into current behaviors and approaches to task completion, dealing with problems, workarounds, and perspectives on competitor products. User research for the purpose of requirements elicitation provides insights that help the product team validate, clarify, or correct assumptions about user needs, improving the relevance and quality of requirements as well as sparking conversations on how to meet requirements. Including people with disabilities in these user research activities produces quality insights to inform accessibility requirements.

Methods for gathering valuable insights into requirements include one-to-one interviews, focus groups and surveys, and contextual inquiry structured around the observation over time of people interacting with a current or similar product. Take steps to include people with disabilities among participants and make sure that the chosen research method is applied in an accessible and inclusive way.[3] Remember that people with disabilities are potential users of a digital product for similar reasons to other potential users, so avoid treating disabled research participants as a separate group only for the purpose of informing accessibility requirements. User research with disabled people plays an important role in an iterative process of designing and developing accessible products, so in Chapter 11, *Testing and Evaluation*, we'll return to the role of user research as an evaluation activity, gathering data to establish how well what has been built meets user accessibility needs.

Do research with users

UX specialist Jared Spool wrote a helpful article outlining different approaches to making design decisions, the extent to which each approach involves conducting user research, and the likelihood that the approach will lead to a product that meets user needs. He defined five decision-making approaches:

1. **Unintended:** Unintended design happens when a team makes no conscious effort to be deliberate in decision-making. (There may not even be a team, as such!) There is no research, and designs happen without any real explanation of why specific functionality or appearance characteristics were chosen, beyond being based on the first or easiest thing the team thought of. Whether this design meets the needs of users is purely down to chance.
2. **Self:** Self-design happens when individuals and teams design according to their own needs. The "research" that is done is based on their own prior experiences. They may assume that they speak for a wider audience, so there is no need to consult beyond the design team. The chances of the design being successful for the team are high, but whether it is usable to a wider audience depends on how representative that team is of the wider audience.
3. **Genius:** Genius design extends self-design to a greater level of success due to the experience and expertise of the team. When decisions are made that are based on years of knowledge derived from research and observation, a highly usable product can result, even without additional research for this specific design challenge. This approach requires a team with many years of experience to be in place, which may be challenging to assemble—there aren't that many geniuses around! And past experience needs to be relevant to the specific characteristics of a new design challenge if genius design is to be successful.
4. **Activity-Focused:** With activity-focused design, teams perform deliberate research activities focused on the tasks that the product is intended to support. Teams will attempt to understand how tasks are best performed and design a user interface around this understanding. However, any user involvement is most likely limited to evaluation of solutions the team has built, influenced by their own research.
5. **User-Focused:** With user-focused design, the team's research focuses on a holistic, contextual understanding of the product's intended audience. There is a deliberate effort to gather data on user characteristics, needs, and wants, and to use this data to inform design decisions.

Importantly, Spool noted that the most effective design teams may adopt different design approaches for different situations, depending on project characteristics, and still achieve success. You can't assume that you're part of a team of geniuses who instinctively know what users with disabilities need, so a user-focused approach is especially likely to lead to a quality user experience for the target audience—and that includes people with accessibility needs.[4]

8.3.3 System Constraints

When figuring out how to implement functional and non-functional accessibility requirements, system constraints may influence how these requirements are met. For example, for closed systems (as discussed in Chapter 4, *Assistive Technology*), there may be limitations in input and output devices, platform accessibility support, and limited capacity for users to apply their own assistive technology. Devices such as kiosks, and other public access terminals must be operable and accessible out-of-the-box, which means providing the necessary accessibility adaptations as part of the hardware and software interface. This context may require embedded hardware and software to provide the necessary accessibility support for people with accessibility needs.

If you're building a digital product that runs on a closed system, you'll need to provide assistive technology options as functionality in a way that you wouldn't need to on an open system, which has implications for the hardware capabilities of the system. Table 8.2 shows how approaches to addressing accessibility needs, influenced by functional performance criteria, vary between open and closed functionality systems and the functional accessibility requirements that may be needed as a result for a closed functionality system.

Table 8.2 Accessibility Needs for Open and Closed Functionality Systems

Accessibility Need	Open Functionality	Closed Functionality
Operation without vision	A user can run their own screen reading software, configured to their preferences. A user can attach a braille display device to receive tactile output, and input content using braille.	The system must be provided with speech output capability, either enabled by default or enabled when the user requests it, for example, when plugging headphones into the system. Physical input and output devices, such as keypads, buttons, card readers, and scanners, must have tactile labels that communicate their purpose.
Operation with limited vision	A user can run their own screen magnification system, adjust display using operating system settings, magnify text using software settings, or adjust display preferences using operating system or software settings.	The system must be provided with a way to ensure that on-screen content is readable, for example, by choosing text of a sufficiently large size, and in a color that contrasts sufficiently with its background. Alternatively, the system must provide options to increase text size or change color settings.

(Continued)

Table 8.2 (*Continued*) Accessibility Needs for Open and Closed Functionality Systems

Accessibility Need	Open Functionality	Closed Functionality
Operation with limited color perception	The system must be designed so that information is presented in a way that does not rely on color perception	Same requirements as for open functionality systems
Operation without hearing	A user can turn on captions in their media player, or have captions set to show by default. At the operating system level, audio alerts can be provided with visual equivalents.	Video must be provided with open captions, or an option to enable captions. The system must not rely on audio as a way to notify users of events.
Operation with limited hearing	A user can adjust audio settings such as volume and background noise control, set audio output to monaural instead of stereo, and choose audio output preferences, for example, speakers or headphones.	The system must be provided with hardware or software options to control volume, provide an option to enable hearing through headphones, or present audio above a minimum decibel value.
Operation without speech	For ICT that accepts vocal input, receiving input through another input channel must be possible, such as a keyboard or touchscreen.	Same requirements as for open functionality systems
Operation with limited manipulation	A user can use their own preferred input device to interact with the system, controlled by gaze, speech, breath, or other means, or make adjustments to how a mouse or keyboard works to reduce effort required to type or manipulate a pointer.	Physical and touchscreen buttons, controls, and other physical devices must be designed in ways that ensure they're operable without requiring fine motor control.

(Continued)

Table 8.2 (*Continued*) Accessibility Needs for Open and Closed Functionality Systems

Accessibility Need	Open Functionality	Closed Functionality
Operation with limited reach and strength	A user can use their own preferred input device to interact with the system, controlled by gaze, speech, breath, or other means.	The user interface must be positioned in a place that allows users in wheelchairs or scooters, or of limited height, to be able to reach and operate controls and read displays while maintaining privacy. Physical buttons, controls, and other input and output devices must be operable without excess force.
Operation with limited language, cognitive and learning abilities	A user can run assistive technology that can read out selected text, highlight text as it is read, or change the display of text to make it easier to read and process.	The system must present text using language understandable to and readable by the target audience, and presented in a way that supports easy reading. The system must carefully manage time limited contents in such a way that users have sufficient time to read text and respond to any prompts, while also meeting security requirements that limit display of sensitive information.
Operation without exposure to photosensitive seizure triggers	Solutions for the user to suppress content that triggers photosensitive seizures are limited at the time of writing. Instead, responsibility lies with content producers or broadcasters of video or animation that includes flickering content to either remove flickering or flashing from the content before publishing, or providing a warning before the user has a chance to be exposed to the flickering content.	Same requirements as for open functionality systems

The accessibility of closed functionality systems has historically lagged behind personal devices that have open functionality. However, there has been progress, in part driven by legislative requirements to ensure that people with disabilities have equal access to devices. For example, many Automated Teller Machines (ATMs) and other kiosks now provide built-in text-to-speech capability, which is triggered when a user inserts headphones into the ATM's headphone jack. This is important, as it meets the goal of enabling blind and visually impaired people to receive audio output from the ATM while also maintaining privacy and security requirements that would be compromised if the speech output was audible to anyone in the vicinity of the device.

For closed functionality systems to be optimally accessible, there needs to be a close partnership between software engineering and industrial design to ensure that the hardware and software work together to meet the necessary accessibility requirements. In the situation of a walk-up-and-use terminal, it's critical to ensure that any accessibility features are clearly identified and easy to operate. Consistency with other similar devices helps the user experience by reducing the time required for a user to figure out how to enable the accessibility features they need.

For open and closed systems, the context of use may also influence how accessibility requirements are expressed. For example, a digital game's purpose is to entertain users, providing a sufficient level of engagement for different skill levels. This may appear to place severe constraints on meeting accessibility requirements. But with a clear definition of the intended gaming experience and a commitment to examining the validity of assumptions about the sensory, physical, or cognitive capability needed to play the game, you can stimulate conversations about how to provide an inclusive gaming experience.

8.4 Requirements Specification

A critical aspect of accessibility requirements is effectively communicating details of the requirements so that all stakeholders in a digital product maintain a shared understanding of product accessibility goals and can ascertain at any given time whether those goals have been met. Without deliberate attention to detail, it can be easy to express accessibility requirements in ambiguous or opaque terms that make it difficult to implement requirements or determine whether they've been met. Accessibility requirements must be managed when product requirements change, for example, when new user insights emerge or business strategy changes. Although core user accessibility needs should remain stable, the emergence of new functional requirements or a change in the development approach or technology platform can impact accessibility requirements.

With the emergence of agile development processes that prioritize functional code over comprehensive documentation, the method of defining complex requirement specifications before starting to develop code has been replaced by a leaner approach to specifying and managing requirements as development progresses. While agile processes may still use definitions of functional requirements and non-functional requirements, common approaches to defining a product's requirements focus on user stories and acceptance criteria. Effectively working accessibility into each of these methods helps reduce the chances that accessibility is neglected during design and development activities, and supports tracking progress toward meeting accessibility requirements through the product development process.

User stories present functional requirements from the perspective of a user with a specific goal in mind. One approach to user stories uses the following format:

As a [type of user] I want to [perform some task] so that I can [achieve some goal].

This approach could be used to articulate accessibility requirements, for example:

As a **screen reader user**, I want to hear **a text alternative for each image providing information** so that I can **understand the information provided by the image**.

In this way, user stories can highlight requirements to meet specific user accessibility needs using a recognizable form, communicating the purpose of accessibility requirements in terms of benefits to users. The shortcoming of expressing accessibility requirements in user stories is that accessibility requirements apply to every piece of functionality to be implemented, so there is a potential scalability issue. Additionally, people with disabilities ultimately have the same set of goals when using the product as other users, and segmenting users with accessibility needs in user stories may incorrectly imply their goals are only to interact with accessibility features.

Acceptance criteria focus on tests that need to be met for functionality to be considered complete. Accessibility acceptance criteria, therefore, communicate to the product team what accessibility tests need to be satisfied for a specific piece of functionality to be considered to have met accessibility requirements. Accessibility acceptance criteria could be expressed as a set of technical accessibility tests that need to be met based on accessibility standards such as WCAG. These tests could be conducted through automated or manual testing methods. Acceptance criteria could also be expressed as performance metrics collected from usability testing with disabled people, which shifts focus to establishing whether the functionality is sufficiently usable. Resources are available to demonstrate how accessibility acceptance criteria can be generated for functionality.[5] Related accessibility requirements could also be incorporated into a shared "definition of done," articulating the core accessibility requirements expected of every piece of functionality built.

Quality accessibility requirements are expressed in ways in which it can be clearly established whether, for a given piece of technology, the requirement has been met. We cover approaches for testing for accessibility in detail in Chapter 11, *Testing and Evaluation*. For now, it's worth emphasizing that accessibility standards such as WCAG have been developed with testability in mind, which makes incorporating accessibility standards into requirements specification and testing more effective. In the next chapter, *Core Requirements*, we'll focus in more depth on how the core accessibility requirements presented in WCAG can be met in the design and implementation stages of the product development process.

Takeaways

As an engineer, you should:

- Understand the role of technical standards in requirements engineering and use established accessibility standards as a starting point for specifying accessibility requirements.
- Recognize the limitations of accessibility standards as the only source for requirements; seek out insights from target users who have accessibility needs to inform accessibility requirements.
- Fully explore the range of business requirements and user requirements related to accessibility when eliciting accessibility requirements and account for them in the requirements specification phase.
- Document accessibility requirements using the method that best integrates with your team's development methodology.

Notes

1 I. Sommerville (2016) *Software Engineering*. 10th Edition. Boston: Pearson Education Limited, p. 105.
2 W3C Note (2021) *Inaccessibility of CAPTCHA: Alternatives to Visual Turing Tests on the Web*. www.w3.org/TR/turingtest
3 D. Aidley and K. Fearon (2021) *Doing Accessible Social Research: A Practical Guide*. Bristol, UK: Policy Press.
4 J. Spool (2009) *5 Design Decision Styles. What's Yours?* articles.uie.com/five_design_decision_styles
5 MagentaA11y *Accessibility Acceptance Criteria*. www.magentaa11y.com

9

CORE REQUIREMENTS

Objectives

Our objectives for this chapter are to present core accessibility requirements based on the Web Content Accessibility Guidelines. We use the WCAG Principles and Guidelines as a framework to explore accessibility requirements. For each guideline, we summarize the associated requirements, focusing on their intent, and discuss how they help address user accessibility needs. We examine considerations and suggested approaches to design, development, and content for implementing core requirements.

Once you're through this chapter, you should:

- Understand fundamental accessibility requirements and the reasoning and intent behind each requirement.
- Be familiar with a range of considerations around designing and implementing core requirements.
- Be ready to apply different approaches to meeting core requirements in your design, development, and content activities.

Introduction

The Web Content Accessibility Guidelines (WCAG) provide us with a well-established and robust set of core accessibility requirements for digital content and functionality that, while prioritized for the web, can be applied to other digital platforms and content types. As such, it's a helpful resource on which to build a core understanding of accessibility requirements. But always keep in mind that WCAG provides a baseline—it is not an exhaustive set of accessibility requirements. For a given resource with a given purpose, there will be other accessibility requirements that are essential for the resource to be usable by people with accessibility needs.

DOI: 10.1201/9781003288060-11

Here, we look more closely at the constituent parts of WCAG 2.2 and the requirements they define, organized in a way that mirrors the structure of WCAG. We restate each WCAG principle, summarize the requirements of each constituent guideline, and outline approaches to meeting requirements in design, content creation, and implementation. We list the specific Level A and Level AA success criteria for each guideline, which at the time of writing is the most commonly referenced conformance level specified in policy, law, and regulations. While the goal of this chapter is to summarize the intent and requirements of each guideline, you should always consult the official W3C reference to ensure that you fully understand each success criterion to establish how it can be satisfied for a specific digital product.

9.1 Perceivable

Information and user interface components must be presentable to users in ways they can perceive.

This broad accessibility principle is broken down into guidelines and requirements for content to be presented such that it can be perceived in multiple sensory channels and can be adapted within a specific sensory channel or context. The requirements also support clear separation of content and presentation, and careful use of color.

Perceivability guidelines and requirements aim at ensuring visual and audio content and functionality are provided in a format that can be consumed by people who have difficulty or are unable to perceive content through those sensory channels. Requirements focus on images, audio, and video content and the need to provide accessible alternatives, including text, captions and transcripts, and audio description. Requirements also focus on optimizing digital products to support alternative ways of experiencing content and interaction. They promote solid structure, simple design, and legible content that people can adapt according to their needs and preferences. Meeting perceivability requirements helps avoid barriers by ensuring people can consume content and functionality in whatever ways best address their accessibility needs.

9.1.1 Text Alternatives

Provide text alternatives for any non-text content so that it can be changed into other forms people need, such as large print, braille, speech, symbols, or simpler language.

This requirement applies to static non-text content—primarily images. Non-text content is not accessible to some people, so the information non-text content conveys must also be available in an accessible format. By providing the equivalent information as a text-based alternative, the information in text can be perceived in different ways, including with

Table 9.1 WCAG 2.2 Requirements for Guideline 1.1—Text Alternatives

Name and Conformance Level	Description
1.1.1 Non-text Content (Level A)	All non-text content that is presented to the user has a text alternative that serves the equivalent purpose, except for the situations listed below.
	• **Controls, Input:** If non-text content is a control or accepts user input, then it has a name that describes its purpose. (Refer to Success Criterion 4.1.2 for additional requirements for controls and content that accepts user input.)
	• **Time-Based Media:** If non-text content is time-based media, then text alternatives at least provide descriptive identification of the non-text content. (Refer to Guideline 1.2 for additional requirements for media.)
	• **Test:** If non-text content is a test or exercise that would be invalid if presented in text, then text alternatives at least provide descriptive identification of the non-text content.
	• **Sensory:** If non-text content is primarily intended to create a specific sensory experience, then text alternatives at least provide descriptive identification of the non-text content.
	• **CAPTCHA:** If the purpose of non-text content is to confirm that content is being accessed by a person rather than a computer, then text alternatives that identify and describe the purpose of the non-text content are provided, and alternative forms of CAPTCHA using output modes for different types of sensory perception are provided to accommodate different disabilities.
	• **Decoration, Formatting, Invisible:** If non-text content is pure decoration, is used only for visual formatting, or is not presented to users, then it is implemented in a way that it can be ignored by assistive technology.

the aid of an assistive technology. Table 9.1 lists the constituent success criteria for this WCAG guideline.

9.1.1.1 Providing Alternatives for Static Content (e.g., Images) Accessibility requirements for text alternatives focus on any static content that is not text, including images of all types, such as photographs, maps, graphs, charts, diagrams, icons that may be used to communicate information, logos, emojis, and decorative imagery. Static content also includes text characters that are used to present graphical content rather than intended to be read as text, such as text-based emoji and ASCII art.

All static non-text must be provided with an equivalent text alternative. In every case, the basic approach is the same: to provide an equivalent text alternative to the information provided by the non-text content. What makes a text alternative *equivalent* depends on the role and purpose of the image. The task of specifying an appropriate text alternative can become subjective, but there are some helpful rules of thumb to follow:

- If an image provides information, then the text alternative should communicate the same information. This might be simple information, such as the logo of an organization or a simple line graph indicating a trend. It might be more complex information, such as a map or pie chart. The more complex the image, the more complex the text alternative needs to be.
- If an image provides a label for an interactive control, such as a link or button, the text alternative should communicate the purpose of the control. For example, the appropriate text alternative for a triangle icon labeling a "play" button on a media player is "Play" and not "Right-pointing triangle."
- For purely decorative images that do not convey information, a text alternative may be pointless and unhelpful. In these cases, the appropriate approach is to hide the image from assistive technology. Different technologies have different methods for hiding images. For example, HTML allows for a null value for the `alt` attribute, and `alt=""` is recognized by assistive technology as a directive to skip the image.

Deciding whether an image is decorative can be subjective. On the one hand, hiding decorative images helps reduce unnecessary clutter. On the other hand, doing so denies users content that may contribute to the experience of using the digital product. For example, a seemingly decorative image may in fact help to communicate that an article is humorous. If an image is what accessibility advocate Léonie Watson refers to as "emotion-rich,"[1] then it's worth trying to communicate that information as a text alternative.

Provisioning of alternatives is an essential part of content and design. Whenever you are preparing content with images, whether for an interface, webpage, or a report or presentation, take the time to provide alternative text that describes the image and provide the alternative using the functionality provided by the platform or format. When creating content and designs for implementation, pay attention to images and text alternatives,

and provide the relevant descriptive details in annotations and requirements. Also, identify and flag all decorative images. Examine the context
for each image and determine an alternative that best describes the image
in context. For complex images where a text alternative might require a
significant amount of text, identify whether that text exists already or
whether it needs to be written. For example, a bar graph is a visual representation of data. If the raw data is available as a table, the table can be a
helpful alternative, along with a text summary of the trend the data is indicating. An image caption can sometimes provide the necessary details for
a text alternative. If an image provides information that is already present
in adjacent text, then mark the image as decorative.

Wherever possible, use native platform methods for programmatically
providing the text alternative as a property of the image, so the alternative is embedded in code with the non-text content. Use the approach
that is appropriate for the coding language and platform. The approach
may vary from a text alternative specified as an attribute of the image to
separate content that is programmatically identified as being a text alternative for the image. For example, for web content, the `alt` attribute of
the `<img>` element and the `<title>` element within the `<svg>` element
both provide ways to specify a text alternative for images. For images in
native mobile apps, the Android `contentDescription` property and iOS
`accessibilityLabel` property serve the same purpose.[2] Also, use the
platform's most appropriate method for specifying decorative images, so
that they are hidden from assistive technology.

9.1.2 Time-Based Media

Provide alternatives for time-based media.

This guideline applies to live and prerecorded video and audio—content
that has a time-based aspect to it. Content types that are most commonly
in scope for time-based media requirements include audio-only content
(such as a podcast recording), video with no audio (silent video), and video
with audio that may include one or more of spoken dialog, music, and
other sounds. Since the logistical challenges of providing real-time equivalent alternatives for time-based media can be significant, formal requirements for equivalent alternatives for live audio and video tend to be less
onerous than for prerecorded audio and video. Table 9.2 lists the constituent success criteria for this WCAG guideline.

9.1.2.1 Providing Alternatives for Audio Accessibility requirements for
audio content focus on providing an alternative for people who do not
have access to the aural channel. Alternatives to audio can be provided
through the following means:

Table 9.2 WCAG 2.2 Requirements for Guideline 1.2—Time-Based Media

Name and Conformance Level	Description
1.2.1 Audio-only and Video-only (Prerecorded) (Level A)	For prerecorded audio-only and prerecorded video-only media, the following are true, except when the audio or video is a media alternative for text and is clearly labeled as such: • **Prerecorded Audio-only:** An alternative for time-based media is provided that presents equivalent information for prerecorded audio-only content. • **Prerecorded Video-only:** Either an alternative for time-based media or an audio track is provided that presents equivalent information for prerecorded video-only content.
1.2.2 Captions (Prerecorded) (Level A)	Captions are provided for all prerecorded audio content in synchronized media, except when the media is a media alternative for text and is clearly labeled as such.
1.2.3 Audio Description or Media Alternative (Prerecorded) (Level A)	An alternative for time-based media or audio description of the prerecorded video content is provided for synchronized media, except when the media is a media alternative for text and is clearly labeled as such.
1.2.4 Captions (Live) (Level AA)	Captions are provided for all live audio content in synchronized media.
1.2.5 Audio Description (Prerecorded) (Level AA)	Audio description is provided for all prerecorded video content in synchronized media.

- **Captions** provide synchronized text that replicates spoken content and non-spoken audio that is critical to understanding a piece of video. Captions are displayed onscreen, superimposed on the video content, and dynamically updated as the video plays. Captions may be closed or open. Closed captions are provided as a separate file and available for a viewer to turn on or off through video player settings when watching the video. Open captions are provided as part of the video content itself, so they can't be turned off.
- A **transcript** is a text alternative to all spoken content in a video or audio recording, provided as a separate text document intended for reading in place of watching a video or listening to an audio recording. While a transcript has the disadvantage of not being synchronized with video or audio as it plays, it can be accessed and reviewed

without the aid of a media player. From an authoring perspective, creating a transcript file avoids the technical challenge of generating timestamp information and creating a caption file.

- **Sign language translation** also provides an alternative to the audio channel, in this case for people whose primary language is sign language. Sign language translations are typically provided as a separate video feed, synchronized, and presented picture-in-picture along with the main video. For live events, a sign language interpreter may be on screen along with the main speakers.

It's worth noting that a significant overlap exists between captions as an accessibility alternative for people who do not have access to the audio channel and text equivalents for people who can hear but have difficulty understanding the audio channel. The latter situation is most commonly found when providing alternatives for dialog in a language not spoken by the audience and is often referred to as "subtitles"—though be aware that in some countries, like the UK, the term subtitles is also used to mean captions. Subtitles focus on translating dialog from one language to another and typically do not include non-spoken audio. The presence of subtitles may be intermittent—only shown when a speaker uses a language different from the primary language of a video's audio track—or persistent, when a recording is provided with subtitles for an international audience. The distinction between the needs of the two groups served by text equivalents means broadcasters, online streaming services, and other providers of video content will typically offer multiple options, including captions in the same language as the video plus subtitles translating the video into other languages.

When creating alternatives for audio content, look for ways to optimize efficiency in creating captions. For example, start with a pre-existing script as a basis for creating a transcript and captions. Make sure to budget time for editing and correcting captions, especially when using automated captioning technology to generate captions. Although automated technology can greatly help with producing accurate timestamps for captions, the speech-recognition algorithms used to generate text from audio may leave errors that must be fixed to produce accurate captions.

Skilled captioners are essential for real-time captioning of live audio and video content. Working with accessible media specialists is a worthwhile investment. Outsourcing captioning to a specialist organization that can quickly turn around accurate captions saves the time and effort of correcting transcription errors. Similarly, engaging specialist sign language interpreters is the best way to create videos with signed translations of audio content.

In addition to providing captions, ensure the media player used to deliver video with audio supports caption display and allows users to customize the display of captions to suit different reading needs, such as changing text size and color.

9.1.2.2 Providing Alternatives for Video Accessibility requirements for video content focus on ways to provide visual content presented in video in a way that's accessible to nonvisual users. This usually means providing an alternative to the visual content for people who do not have access to the visual channel.

- Ensure that the spoken dialog covers all information conveyed visually in the video, sometimes referred to as self-describing video.
- Provide an additional audio channel that contributes a spoken description of visual events not apparent from the main audio track. This is often referred to as an audio description and sometimes as a video description or descriptive narration.
- If the video has no audio, then a fallback option is to provide a media transcript with information and important details from the video provided as text.

When creating alternatives for video, explore ways that video scripts can encourage self-described video, where the speaker dialog describes visual scenes and information in a way that conveys the equivalent meaning in the spoken audio. Provide quality alternatives with accurate captions and an audio description track with enough detail to convey important visual events and information to nonvisual users.

9.1.3 Adaptable

Create content that can be presented in different ways (for example, simpler layout) without losing information or structure.

This guideline addresses the need to design and present content in a way that is adaptable to different sensory channels and accommodates different needs within a specific sensory channel. It recognizes that people with accessibility needs perceive content in different ways, particularly through sight, hearing, or touch, and may have approaches that adapt how they perceive content, for example, zooming the display or changing colors. Table 9.3 lists the constituent success criteria for this WCAG guideline.

9.1.3.1 Providing Information About Semantics, Structure, and Relationships Accessibility requirements for providing information about semantics, structure, and relationships focus on ensuring the meaning and structure of content are encoded to the greatest extent possible in

Table 9.3 WCAG 2.2 Requirements for Guideline 1.3—Adaptable

Name and Conformance Level	Description
1.3.1 Info and Relationships (Level A)	Information, structure, and relationships conveyed through presentation can be programmatically determined or are available in text.
1.3.2 Meaningful Sequence (Level A)	When the sequence in which content is presented affects its meaning, a correct reading sequence can be programmatically determined.
1.3.3 Sensory Characteristics (Level A)	Instructions provided for understanding and operating content do not rely solely on sensory characteristics of components such as shape, color, size, visual location, orientation, or sound.
1.3.4 Orientation (Level AA)	Content does not restrict its view and operation to a single display orientation, such as portrait or landscape, unless a specific display orientation is essential.
1.3.5 Identify Input Purpose (Level AA)	The purpose of each input field collecting information about the user can be programmatically determined when: • The input field serves a purpose identified in the Input Purposes for User Interface Components section; and • The content is implemented using technologies with support for identifying the expected meaning for form input data.

digital products. In this way, information about meaning and structure is available to assistive technology and available to users in a way that meets their accessibility needs.

Design techniques use many different presentational attributes to communicate the semantic meaning and structure of content. Text can vary in characteristics such as typeface, size, weight, color, background, and border. These characteristics communicate what that text is and provide a hierarchical indication of its importance. The appearance and relative position of groups or blocks of content communicate that objects in the group are related (or not) and that the group has relationships with other groups. When information about content that is expressed visually is also expressed in code, assistive technologies can communicate the same information about meaning, structure, and relationships in alternative ways.

In markup languages such as HTML, this means using the correct elements to identify a piece of content's semantic meaning—for example, whether it's a heading, paragraph of text, list item, quote, or a column header in a data table. It also means identifying groups of related content. For example, a collection of list items can be grouped and semantically identified as an ordered or unordered list. For data tables, table cells are grouped in table rows and columns, which in turn are grouped in a table with column and row headers.

On web pages, content groups can be identified at a page level as regions or landmarks. HTML 5 region elements provide a way to communicate structure, but for some older user agents, they may not have full support. As a fallback, WAI-ARIA provides a set of role attribute values that can be used to specify landmarks, similar to regions.[3] For example, the main content area can be identified as such in HTML, while a group of content that serves as a footer region can be identified as a footer. With semantic markup, the key is to provide information that correctly expresses the purpose and structure of elements rather than the visual representation of the content. Misusing semantics elements, such as marking up text that doesn't structurally serve as a heading in order to achieve large, bold text, can cause issues for users who rely on encoded semantics to navigate document structure.

Approaches to providing information about semantics, structure, and relationships include the following:

- Identify the semantic meaning of each piece of content and relationships within content, for example, lists, tables, groups of input elements, and labels and error messages for input elements in a visual design.
- Annotate semantic information in designs in a way that development can reflect this information in code, including identifying discrete regions and groups of content and relationships between content, such as between error messages and the source of the error.
- Use the correct semantic elements to implement designs, such as correct HTML elements and attributes for web content, to programmatically identify the specified semantic meaning of each piece of content, like headings and regions, and to connect related pairs or groups of content. Examples of pairs of content items that need to be associated in code include an error message and source of the error, and a label and an input field. Groups of content items that need to be identified programmatically as a group include a set of radio button options and checkbox options.
- When identifying regions of content on a web page, make sure that every piece of content on the page exists in a container element that is identified as a region to avoid orphaned content that may be missed by an assistive technology user navigating from region to region.

- Provide text in appropriately structured content, using headings, lists, quotes, and tables to structure content.

9.1.3.2 Sequencing Content for Meaning The intended sequence in which content should be read must be communicated in code so that content can be presented in the correct order by assistive technology. This requirement addresses the need for content to be successfully linearized from a two-dimensional layout to a single ordered sequence of content that makes sense, for example, when read out by a screen reader. This requirement particularly addresses situations where content is presented in multi-column layouts and where content flows around boxes of other content.

Approaches to providing a meaningful sequence of content include the following:

- When creating a static visual design of a user interface, clearly annotate the intended sequence of content when linearized, for example, when read by a screen reader or presented in a single-column text display. This helps communicate to developers in what order content should be arranged.
- Carefully design the behavior of content intended to be hidden in certain interface states so that its presence or absence is logical for all users.
- When implementing content, ensure as far as possible that the code order of the content matches the preferred reading order and the linear order of the content in the interface. This means being cautious about using any technique that visually positions content in a different order to code order, such as CSS float techniques in web design.
- When content is intended to be hidden from users, make sure that techniques for hiding it visually are also implemented programmatically to ensure the content is hidden from assistive technology. In web design, positioning content off screen may leave it reachable by a screen reader user, so take care to use a reliable means of hiding content, such as the CSS `display:none` property. Similarly, where content is visually constrained within a specific interface, such as a modal dialog, make sure those constraints also apply to the screen reader experience.
- In closed functionality technology where a custom text-to-speech solution is provided, specify the text-to-speech output in a way that allows users to correlate the speech output with the information presented on screen.

9.1.3.3 Avoiding Reliance on Sensory Characteristics When interfaces provide instructions that help users understand and interact with content, instructions must not rely on sensory characteristics for understanding. This requirement specifically addresses the need to ensure that text or spoken instructions don't assume that users can distinguish specific sensory characteristics, such as shape, position, size, or sound. Problems arise if instructions refer to interface elements only by a sensory characteristic. Note that this requirement applies to instructions, not to the design of the interface element they apply to. It's perfectly reasonable to use sensory characteristics such as size and relative position on screen to help distinguish elements.

Consider the following examples of an instruction from a quiz interface. The first example does not meet the requirement because the instruction relies on the visual perception of shape, color, and position on the page.

> *To give up without answering, click the round red button on the right-hand side of the question page.*

The next example meets the requirement because the instruction includes a reference to the button's name. Even if the button's name isn't visible, as long as the name is announced by screen readers, then the instruction makes sense to people who don't see it.

> *To give up without answering, click the round "Quit" button on the right-hand side of the question page.*

When writing instructions for a user interface, review the content for any reference to something that can only be perceived through one sense, for example, vision, hearing, or touch. Where you do find such content, check whether the instructions could be understood by someone without access to that sensory channel.

9.1.3.4 Adapting to Different Device Orientations When viewed on a device that can change orientation between portrait and landscape, content must adapt to each orientation unless a specific orientation is essential to the purpose of the content. This requirement addresses the needs of users who require a fixed orientation when using a smartphone, tablet, or other device that can change display orientation. For example, some people with a physical disability that affects motor control may use a device mounted in a fixed orientation on a wheelchair arm; some people with low vision may prefer a particular orientation for ease of reading. When content in a left-to-right language like English does not adapt to a device's orientation, reading may become very difficult due to text being oriented vertically; content may also be missing or obscured.

The majority of interfaces should be adaptable to being viewed in portrait and landscape mode without loss of functionality, even if one orientation may provide a more optimized user experience. Rare exceptions to this requirement may be specific situations where a particular orientation is essential for the experience, potentially including certain games or where the device technology cannot change orientation, such as a virtual reality application experienced using a headset with a fixed orientation screen.

Approaches to adapting to different device orientations include the following:

- Avoiding using code that locks the interface into a fixed orientation, for example, the CSS `orientation` media query.
- Designing interface flexibility so that layout is optimized for different orientations. This may include repositioning interface controls to be easier to operate and adjusting the placement of content.

9.1.3.5 Supporting Form Fill-In   Where an input field collects specific types of information about a user, the field's input purpose must be programmatically determined. This requirement helps make form-filling easier for people who might otherwise have difficulties completing forms due to conditions that affect memory, language, or communication, or physical disabilities that make inputting content more difficult. Identifying input purpose programmatically can help users by supporting autofill functionality provided by platforms such as a browser and can support smartphones and tablets in customizing onscreen keyboards for input type. Communicating the input purpose programmatically also provides the potential for fields to be supplemented by symbols or images that represent the purpose of the input field in recognizable ways to people who use symbol-based communication systems.

When implementing form input fields that collect specific items of information about a user, ensure that the input purpose is specified in the code. WCAG defines a specific set of personal information attributes as being in scope for this requirement in the section "Input Purposes for User Interface Components."[4] Examples include given name, family name, address fields, date of birth, password, and personal payment card details. The basic application of this requirement is restricted to input fields that collect data about a user. If a form is collecting data that may not be personal to the user, then it is not subject to this requirement. For example, in an online shop's checkout form, the name and address fields for recipient details might not be the same as the purchaser if the item is a gift, so they can't be considered personal information.

9.1.4 Distinguishable

Make it easier for users to see and hear content including separating foreground from background.

This guideline presents a wide-ranging set of requirements addressing the need to provide content in a way that allows users to adapt how that content is presented in a specific sensory channel. It recognizes the importance of content legibility and optimizes designs and interactions to foreground and define the most important content and functionality Table 9.4 lists the constituent success criteria for this WCAG guideline.

Table 9.4 WCAG 2.2 Requirements for Guideline 1.4—Distinguishable

Name and Conformance Level	Description
1.4.1 Use of Color (Level A)	Color is not used as the only visual means of conveying information, indicating an action, prompting a response, or distinguishing a visual element.
1.4.2 Audio Control (Level A)	If any audio on a Web page plays automatically for more than 3 seconds, either a mechanism is available to pause or stop the audio, or a mechanism is available to control audio volume independently from the overall system volume level.
1.4.3 Contrast (Minimum) (Level AA)	The visual presentation of text and images of text has a contrast ratio of at least 4.5:1, except for the following: • **Large Text:** Large-scale text and images of large-scale text have a contrast ratio of at least 3:1; • **Incidental:** Text or images of text that are part of an inactive user interface component, that are pure decoration, that are not visible to anyone, or that are part of a picture that contains significant other visual content, have no contrast requirement. • **Logotypes:** Text that is part of a logo or brand name has no contrast requirement.
1.4.4 Resize text (Level AA)	Except for captions and images of text, text can be resized without assistive technology up to 200 percent without loss of content or functionality.
1.4.5 Images of Text (Level AA)	If the technologies being used can achieve the visual presentation, text is used to convey information rather than images of text except for the following:

(Continued)

Table 9.4 (*Continued*) WCAG 2.2 Requirements for Guideline 1.4—Distinguishable

Name and Conformance Level	Description
	• **Customizable:** The image of text can be visually customized to the user's requirements; • **Essential:** A particular presentation of text is essential to the information being conveyed.
1.4.10 Reflow (Level AA)	Content can be presented without loss of information or functionality, and without requiring scrolling in two dimensions for: • Vertical scrolling content at a width equivalent to 320 CSS pixels; • Horizontal scrolling content at a height equivalent to 256 CSS pixels; Except for parts of the content which require two-dimensional layout for usage or meaning.
1.4.11 Non-text Contrast (Level AA)	The visual presentation of the following have a contrast ratio of at least 3:1 against adjacent color(s): • **User Interface Components:** Visual information required to identify user interface components and states, except for inactive components or where the appearance of the component is determined by the user agent and not modified by the author; • **Graphical Objects:** Parts of graphics required to understand the content, except when a particular presentation of graphics is essential to the information being conveyed.
1.4.12 Text Spacing (Level AA)	In content implemented using markup languages that support the following text style properties, no loss of content or functionality occurs by setting all of the following and by changing no other style property: • Line height (line spacing) to at least 1.5 times the font size; • Spacing following paragraphs to at least 2 times the font size; • Letter spacing (tracking) to at least 0.12 times the font size; • Word spacing to at least 0.16 times the font size.

(Continued)

Table 9.4 (*Continued*) WCAG 2.2 Requirements for Guideline 1.4—Distinguishable

Name and Conformance Level	Description
	Exception: Human languages and scripts that do not make use of one or more of these text style properties in written text can conform using only the properties that exist for that combination of language and script.
1.4.13 Content on Hover or Focus (Level AA)	Where receiving and then removing pointer hover or keyboard focus triggers additional content to become visible and then hidden, the following are true: • **Dismissible:** A mechanism is available to dismiss the additional content without moving pointer hover or keyboard focus, unless the additional content communicates an input error or does not obscure or replace other content; • **Hoverable:** If pointer hover can trigger the additional content, then the pointer can be moved over the additional content without the additional content disappearing; • **Persistent:** The additional content remains visible until the hover or focus trigger is removed, the user dismisses it, or its information is no longer valid. Exception: The visual presentation of the additional content is controlled by the user agent and is not modified by the author.

9.1.4.1 Avoiding Reliance on Color Perception Color is a powerful way to communicate information and an important feature of an accessible and usable digital product. However, not everyone perceives colors in the same way as intended by a visual designer, so the designer's intended meaning associated with a specific color may be lost. This means that there must be a way to perceive information without relying on color. This requirement specifically benefits users who have a color deficit that affects their perception of or ability to distinguish between specific hues.

Approaches to avoiding reliance on color include:

• Provide the same information communicated by color in text.
• Apply other visual characteristics to content along with color, such as changing font weight or providing a distinct border.
• Ensure that information is presented in a way that does not rely on color perception, for example, in the presentation of error message text or in diagrams and charts used to communicate information.

To avoid designing using color in ways that might result in barriers, consider how the content and functionality would be experienced in a monochrome context without color. If the information provided using color is not discernible without color, use an additional design approach to visually communicate the information.

9.1.4.2 Providing Sufficient Color Contrast Contrast plays a large part in clarity of elements in a design, including text and other elements. Legibility is dependent on the contrast between background elements and foreground elements. Low-contrast color combinations can cause accessibility barriers. People who have low vision may have difficulty distinguishing elements on a screen when they are implemented using colors that lack contrast. This applies to text content as well as non-text content, including control borders, infographics, and icons. If it's important that users are able to perceive a design element or content, that means it's important to all users, including people with vision impairments that affect their ability to perceive color contrast.

When designing color palettes for use in an interface, avoid introducing low-contrast color schemes that reduce the readability of text and significant non-text content. Be sure your color choices do not make it hard for users to distinguish text from its background. WCAG defines minimum contrast ratios that can be used to compare two color values for contrast. Also recognize that some users may need to change colors to make content easier to read, for example, by inverting colors or by reducing contrast. Implement colors in a way that respects different user preferences and settings.

9.1.4.3 Supporting Adjustments to Text Appearance Digital platforms allow users to adjust text appearance to make it easier to read and understand, for example, by enlarging text, adjusting letter and line spacing, and changing color or font using system and software settings. Changes to text appearance can affect other factors in a design, like margins, spacing, and column widths. This requires text content and page and screen layouts to be designed and implemented to support a range of text adjustments.

There are several approaches to supporting adjustments to text appearance, including the following:

- Create designs and layouts that will adapt gracefully to changes in text appearance.
- Avoid implementing designs in a way that restricts adjustments to text appearance. For example, avoid hard-coding text size or the size of containers of text, and avoid where possible using images of text.
- Support text display settings, such as changes to size, color, and different viewing modes, such as increased or high-contrast mode, dark mode, and reader mode in web browsers.

9.1.4.4 Avoiding Disruptive and Obscuring Content There are several requirements that aim to ensure other elements in a design do not get in the way of important content and interaction. For example, audio can be problematic for screen reader users who rely on being able to hear the screen reader output. When a digital product includes audio that auto-plays, the audio from the product interferes with the screen reader's audio. Another requirement addresses issues that arise from content that overlays an interface, for example, a drop-down menu or tooltip that appears on mouse hover and overlays important page content. In both cases, users must manage the disruption itself and also find ways to control the obscuring content so that they can proceed with their task, either by dismissing the content or making effective use of the information it provides.

Approaches to avoiding disruptions include:

- Avoid using auto-playing audio on a digital interface. If for some reason auto-playing audio is required, provide a clear and easily accessible control to stop the audio.
- Avoid excess use of design elements that are not persistent in an interface. If an element is sufficiently important to include, try to find a way to present it as a persistent element. For example, label interface controls with a persistent text label and icon rather than providing the text label on hover.
- For a design element that must be implemented in a non-persistent way, implement the element such that users can control its display and have time to consume its content.

9.2 Operable

User interface components and navigation must be operable.

This broad accessibility principle is broken down into requirements that address various aspects of interface operability, ensuring that people can interact with user interface elements such as links, buttons, menus, and input fields, regardless of interaction modality and accessibility needs.

Operability includes requirements to address the accessibility needs of people who use diverse input methods and devices, including keyboard, mouse, touch, and voice. It also includes requirements to limit negative impacts on users, such as content types that are distracting, obstructive, and harmful. Operability requirements recognize the growing diversity in input devices and methods that people with accessibility needs may encounter, in particular the emergence of touchscreens, device activation, and voice interaction as common input methods. Meeting operability requirements helps ensure people can operate interfaces using different methods without encountering unnecessary barriers or experiencing harm.

9.2.1 Keyboard Accessible
Make all functionality available from a keyboard.

Requirements under this guideline emphasize the importance of ensuring that interaction is possible with a keyboard or keyboard equivalent. These requirements benefit people who, because of a physical, motor, or cognitive disability, do not use a pointer-based input device, such as mouse, or have difficulty using a pointer device to accurately select and activate user interface elements. Table 9.5 lists the constituent success criteria for this WCAG guideline.

9.2.1.1 Supporting Keyboard Interaction Functionality provided by the interface must be operable using a keyboard interface. A keyboard interface may range from a full keyboard to a more limited-input device, such as a keypad used for a kiosk, games console, or TV remote control. Even

Table 9.5 WCAG 2.2 Requirements for Guideline 2.1—Keyboard Accessible

Name and Conformance Level	Description
2.1.1 Keyboard (Level A)	All functionality of the content is operable through a keyboard interface without requiring specific timings for individual keystrokes, except where the underlying function requires input that depends on the path of the user's movement and not just the endpoints.
2.1.2 No Keyboard Trap (Level A)	If keyboard focus can be moved to a component of the page using a keyboard interface, then focus can be moved away from that component using only a keyboard interface, and, if it requires more than unmodified arrow or tab keys or other standard exit methods, the user is advised of the method for moving focus away.
2.1.4 Character Key Shortcuts (Level A)	If a keyboard shortcut is implemented in content using only letter (including upper- and lower-case letters), punctuation, number, or symbol characters, then at least one of the following is true: • **Turn off:** A mechanism is available to turn the shortcut off; • **Remap:** A mechanism is available to remap the shortcut to include one or more non-printable keyboard keys (e.g., Ctrl, Alt); • **Active only on focus:** The keyboard shortcut for a user interface component is only active when that component has focus.

interactions that may seem to require a pointing device or touchscreen gesture, such as drawing or drag-and-drop functionality, should be operable using a keyboard. The exception would be any functionality that depends on a specific path taken by the input device, for example, drawing or handwriting.

Common keyboard keystrokes include:

- TAB to move focus to the next focusable element.
- SHIFT + TAB to move focus to the previous focusable element.
- DOWN ARROW to move focus to the next item in a select menu or list of radio buttons.
- UP ARROW to move focus to the previous item in a select menu or list of radio buttons.
- SPACEBAR to check or uncheck a checkbox or radio button.
- ENTER to activate a button or link.
- ESC to close a dialog or popup menu.

In addition to supporting the operation of an interface using keyboard keys, there are other considerations. Keyboard interaction should not require specific timings to trigger functionality, such as a rapid repeat press of the same key or holding a key down for a certain amount of time. Timing requirements may be difficult or impossible for some users to perform. Additionally, with single-key application-specific keyboard shortcuts, users need to have a way to turn off the shortcut or change it to another key. This functionality helps ensure that application-specific keyboard shortcuts do not interfere with common keyboard navigation and keyboard shortcuts provided by assistive technology.

When keyboard interaction is supported, the interface must not trap keyboard focus on any interactive control. It must be possible to move keyboard focus to and away from every interactive element on the page or screen. Keyboard traps cause problems for keyboard users, who may have to reload the page or stop interacting altogether with certain parts of a page or screen. This requirement exists in response to inadvertent instances of scripts that fail to properly respond to keystrokes and embedded objects in an interface that retain focus.

Ways to support keyboard accessibility include:

- Implement all interactive elements on a page or screen in a way that ensures they are focusable and operable using standard keyboard methods. Do not rely on specific timings of key presses. Avoid using application-specific single-key keyboard shortcuts where possible.

- Design and implement interactions that honor standard keystrokes for moving focus between and activating interactive elements wherever possible. Provide clear instructions where non-standard keystrokes are used to operate interactive elements.
- For any element that has scripted behavior responding to user input events, ensure that keyboard behaviors do not result in a keyboard trap. Make sure the focus can move to the next or previous element.

9.2.2 Enough Time

Provide users enough time to read and use content.

This guideline focuses on requirements that ensure there are ways for users to control time-limited and moving content. Sensory, cognitive, and motor disabilities may create a user accessibility need for more time to read and interact with content. For some users, animations can cause sufficient distraction to make reading and interacting with an interface difficult or impossible. Table 9.6 lists the constituent success criteria for this WCAG guideline.

Table 9.6 WCAG 2.2 Requirements for Guideline 2.2—Enough Time

Name and Conformance Level	Description
2.2.1 Timing Adjustable (Level A)	For each time limit that is set by the content, at least one of the following is true: • **Turn off:** The user is allowed to turn off the time limit before encountering it; or • **Adjust:** The user is allowed to adjust the time limit before encountering it over a wide range that is at least ten times the length of the default setting; or • **Extend:** The user is warned before time expires and given at least 20 seconds to extend the time limit with a simple action (for example, "press the space bar"), and the user is allowed to extend the time limit at least ten times; or • **Real-time Exception:** The time limit is a required part of a real-time event (for example, an auction), and no alternative to the time limit is possible; or • **Essential Exception:** The time limit is essential and extending it would invalidate the activity; or • **20-Hour Exception:** The time limit is longer than 20 hours.

(Continued)

Table 9.6 (*Continued*) WCAG 2.2 Requirements for Guideline 2.2—Enough Time

Name and Conformance Level	Description
2.2.2 Pause, Stop, Hide (Level A)	For moving, blinking, scrolling, or auto-updating information, all of the following are true: • **Moving, blinking, scrolling:** For any moving, blinking or scrolling information that (1) starts automatically, (2) lasts more than five seconds, and (3) is presented in parallel with other content, there is a mechanism for the user to pause, stop, or hide it unless the movement, blinking, or scrolling is part of an activity where it is essential; and • **Auto-updating:** For any auto-updating information that (1) starts automatically and (2) is presented in parallel with other content, there is a mechanism for the user to pause, stop, or hide it or to control the frequency of the update unless the auto-updating is part of an activity where it is essential.

9.2.2.1 Providing Options for Time-Limited Interactions Accessibility requirements for time-limited interactions aim at ensuring users who need more time to complete tasks have the means to extend or turn off time limits. Some time limits on the content may be required. For example, security requirements for an application that supports sensitive data transactions may mandate an automatic logout after a period of user inactivity. When a time limit is provided and required, there must be a way for users to change or extend its duration, unless doing so is impractical due to the nature of the activity, for example, a real-time event like an auction. Where a time limit is present but not required, there must be a way for users to turn off the time limit.

When designing and implementing systems with time limits, design the time-out interaction to support users who need more time. For example, alert users to an approaching time limit through a clear message and provide an option to extend the time limit, for example, by pressing an "OK" button to extend. Make sure the message is presented onscreen for a sufficient time to allow users to notice and respond to the message, and make sure the notification is implemented in a way that ensures assistive technology can detect and announce its presence (see requirements for *Compatibility* later in this chapter).

9.2.2.2 Providing Options to Control Moving Content The aim of accessibility requirements for moving content is to make sure users who are adversely affected by motion and animation have options for controlling the content. Where an interface includes content that moves, blinks, or scrolls, users need a way to pause or stop the movement or hide the moving content. For content that automatically updates, users need the means to pause or stop the updating or to change the update frequency.

Be cautious about including animation in content and interactions, and avoid any unnecessary use of animation. Where an operating system or platform allows users to specify preferences for reduced motion, check for and honor these preferences. In cases where animation is considered essential, either provide animation only for a short time period or provide clearly identifiable and accessible controls that allow users to pause, stop, or hide the animated content. Avoid auto-playing video content wherever possible. For applications that provide users with profile and customization options, provide a settings option in a user's profile, allowing them to turn off or hide animations throughout the application.

9.2.3 Seizures and Physical Reactions

Do not design content in a way that is known to cause seizures or physical reactions.

This guideline specifically focuses on the needs of people who may experience seizures or other physical reactions to flashing and animated content. Given the specific nature of this user accessibility need, exposure to any flashing content is potentially harmful, and no assistive technology yet exists that can suppress flashing content. The only safe approach to take is to design ways to avoid users experiencing this content. Table 9.7 lists the constituent success criteria for this WCAG guideline.

9.2.3.1 Avoiding Flashing and Flickering Content These accessibility requirements provide a defined acceptable level of flashing and flickering and guidelines for staying within those parameters. The requirements include guidance for how to manage content that exceeds the safe threshold to minimize any negative or harmful impacts on people who might be affected.

Table 9.7 WCAG 2.2 Requirements for Guideline 2.3—Seizures and Physical Reactions

Name and Conformance Level	Description
2.3.1 Three Flashes or Below Threshold (Level A)	Web pages do not contain anything that flashes more than three times in any one second period, or the flash is below the general flash and red flash thresholds.

To minimize the potential for seizures and physical reactions, avoid providing any content that flashes or flickers at a frequency of more than three times in a second. In cases where including such content is unavoidable, provide sufficient notification to users who might be affected so that they can avoid encountering the content. For example, for a video that includes flash photography, provide a warning that the content includes flashing on the page that contains the video. Do not auto-play any video or animation that contains flashing content, and always include controls to stop playing flashing content. Include a warning prominently in links to pages that include flashing content so that users can choose not to follow these links.

9.2.4 Navigable

Provide ways to help users navigate, find content, and determine where they are.

Requirements for this guideline address aspects of accessible wayfinding when operating digital interfaces. They focus on accessible ways to support navigation and orientation, including efficient and effective keyboard focus management and indication of focus, clear links, descriptive screen and page titles, section headings, and input labels. They also cover the need to provide flexibility in navigation options. These requirements benefit people who navigate and interact using assistive technology and people who may have difficulty successfully finding their way around a digital product. Table 9.8 lists the constituent success criteria for this WCAG guideline.

Table 9.8 WCAG 2.2 Requirements for Guideline 2.4—Navigable

Name and Conformance Level	Description
2.4.1 Bypass Blocks (Level A)	A mechanism is available to bypass blocks of content that are repeated on multiple Web pages.
2.4.2 Page Titled (Level A)	Web pages have titles that describe topic or purpose.
2.4.3 Focus Order (Level A)	If a Web page can be navigated sequentially and the navigation sequences affect meaning or operation, focusable components receive focus in an order that preserves meaning and operability.
2.4.4 Link Purpose (In Context) (Level A)	The purpose of each link can be determined from the link text alone or from the link text together with its programmatically determined link context, except where the purpose of the link would be ambiguous to users in general.

(Continued)

Table 9.8 (*Continued*) WCAG 2.2 Requirements for Guideline 2.4—Navigable

Name and Conformance Level	Description
2.4.5 Multiple Ways (Level AA)	More than one way is available to locate a Web page within a set of Web pages except where the Web Page is the result of, or a step in, a process.
2.4.6 Headings and Labels (Level AA)	Headings and labels describe topic or purpose.
2.4.7 Focus Visible (Level AA)	Any keyboard operable user interface has a mode of operation where the keyboard focus indicator is visible.
2.4.11 Focus Not Obscured (Minimum) (Level AA)	When a user interface component receives keyboard focus, the component is not entirely hidden due to author-created content.

9.2.4.1 Supporting Different Ways of Navigating People who rely on a keyboard or keyboard-equivalent input device may find the effort of interacting with an interface physically challenging, particularly where the interface has many interactive elements. Keyboard interaction involves moving focus sequentially through links and controls. Reducing the physical effort and time required to operate an interface using a keyboard interface means paying particular attention to how keyboard focus is managed.

Provide ways to bypass blocks of content, particularly blocks of content that are repeated on multiple pages, such as navigation menus that are often found along the top or down the left-hand side of pages and screens. One convention is to provide a link before the start of a set of interactive content that is repeated on multiple pages, allowing keyboard users to move focus to a location immediately after the repeated content. Often referred to as a "skip link," this feature can be implemented to be visibly hidden until it receives keyboard focus.

Make sure that the order of focus through all active elements on a page is logical, so that keyboard operation is as efficient as possible. In most cases, this means ensuring that the focus order matches the visual layout of interactive elements. In most situations, focus order should be limited to elements that are (1) interactive and (2) present on screen. Any static element with no interactivity or that is currently hidden from view should not be focusable. For example, items in a drop-down menu should only be included in focus order when the menu is displayed.

To support the flexibility of navigation between collections of webpages, provide multiple ways to allow access to each page. Alternative navigation methods include providing navigation menus in the header or footer, a search feature, a table of contents, or a site map with links to all pages. This approach supports users who may find a specific navigation method most efficient for their needs.

9.2.4.2 Providing Wayfinding and Orientation Features Requirements for wayfinding and orientation aim to help users recognize where they are on a website, interface, or a specific page or screen, making it easier to understand how content is organized and reducing the likelihood of becoming lost.

One key aspect of wayfinding is a "you are here" indicator. In a digital interface, one key orientation feature is a page or screen title. When a title succinctly describes the topic or purpose of the screen, users know where they are and what they will find there. The title is typically the first thing read out by a screen reader when the page loads, providing immediate orientation information to a screen reader user. Headings are another important orientation indicator, describing the purpose or topic of the content that follows. Think of headings as signposts and use language that clearly describes the content that follows.

For keyboard orientation within page or screen content, make sure there is a clear, visible indicator that allows easy detection of the element that currently has focus. Most platforms provide a native focus indication style, which, in some cases, can be hard to visually distinguish. Include focus styles for links, buttons, and other interactive elements in visual design specifications and apply these styles when implementing interactive elements. Any element receiving focus must not be entirely hidden by other interface content, so when implementing features that remain in a fixed position on a page while scrolling, such as sticky headers and footers, implement them in a way that avoids obscuring the element that currently has keyboard focus. If a hidden element must receive focus, such as a skip link, make sure that the element becomes visible when focused.

Links are an essential wayfinding feature, providing the means to navigate around a digital product. Where links to other pages are provided in the content, make sure that the text of the link indicates the link's purpose or destination, preferably without relying on surrounding text for context. Since links should be styled to be distinct from surrounding text, users may scan content to identify the presence of links. Some assistive technologies, like screen readers, can generate a list of all links on a page, allowing users to quickly search through links for one that may be of interest to their task or goal. Ambiguous link text makes it difficult to quickly determine the destination of the link, reducing the ease of navigation and orientation and potentially increasing navigation errors. For example, consider a webpage about an event with a link to a booking site that uses a single word link, "here":

"In order to ensure you get the chance to experience this fantastic event, click <u>here</u> to book your tickets and make sure you don't miss out on what is sure to be a great occasion!"

The "here" link provides no details about the destination of the link. The only way this link makes sense is in the context of the surrounding text. A simple rewording of the link text clearly indicates the purpose of the link without requiring users to read any additional text:

"<u>Book your tickets for this event</u> – don't miss out on what is sure to be a great occasion!"

Similarly, make sure that each button, form input field, and other interactive elements have a label that clearly describes their purpose. For text input fields, the label's purpose is the data to be collected by the field; for input elements like checkboxes and radio buttons, the purpose is the value of the data represented by the checkbox or radio button.

9.2.5 Input Modalities

Make it easier for users to operate functionality through various inputs beyond keyboard.

This guideline applies the operability principle to input methods other than keyboards with the aim of making pointer-based, speech-based, and touchscreen interactions easier to perform. For example, some people with physical disabilities prefer to use a mouse or touchscreen over a keyboard or keyboard equivalent, but nevertheless find it more difficult to perform pointer and touch interactions that require fine levels of dexterity. Some people use speech as a primary input method but may encounter challenges effectively using speech input if an interface is not optimized to support speech input technology. And some users may use a combination of input devices and methods, changing their device or method according to their specific accessibility needs and their context of use.

Requirements for input modalities support speech input and address issues that arise from using motion and certain pointer interactions for interaction and input. Table 9.9 lists the constituent success criteria for this WCAG guideline.

9.2.5.1 Supporting a Range of Pointer Interactions Supporting a range of pointer interactions involves designing interactions to allow flexible pointer-based input, such as using a mouse pointer, finger, or stylus, and minimize the chances of input errors.

When implementing interactions that can be activated using pointer gestures, support single-click and one-finger gestures. Avoid relying on gestures that require multiple points of contact, such as two or more fingers,

Table 9.9 WCAG 2.2 Requirements for Guideline 2.4—Input Modalities

Name and Conformance Level	Description
2.5.1 Pointer Gestures (Level A)	All functionality that uses multipoint or path-based gestures for operation can be operated with a single pointer without a path-based gesture, unless a multipoint or path-based gesture is essential.
2.5.2 Pointer Cancellation (Level A)	For functionality that can be operated using a single pointer, at least one of the following is true: • **No Down-Event:** The down-event of the pointer is not used to execute any part of the function; • **Abort or Undo:** Completion of the function is on the up-event, and a mechanism is available to abort the function before completion or to undo the function after completion; • **Up Reversal:** The up-event reverses any outcome of the preceding down-event; • **Essential:** Completing the function on the down-event is essential.
2.5.3 Label in Name (Level A)	For user interface components with labels that include text or images of text, the name contains the text that is presented visually.
2.5.4 Motion Actuation (Level A)	Functionality that can be operated by device motion or user motion can also be operated by user interface components and responding to the motion can be disabled to prevent accidental actuation, except when: • **Supported Interface:** The motion is used to operate functionality through an accessibility supported interface; • **Essential:** The motion is essential for the function and doing so would invalidate the activity.
2.5.7 Dragging Movements (Level AA)	All functionality that uses a dragging movement for operation can be achieved by a single pointer without dragging, unless dragging is essential or the functionality is determined by the user agent and not modified by the author.

(Continued)

Table 9.9 (*Continued*) WCAG 2.2 Requirements for Guideline 2.4—Input Modalities

Name and Conformance Level	Description
2.5.8 Target Size (Minimum) (Level AA)	The size of the target for pointer inputs is at least 24 by 24 CSS pixels, except where: • Spacing: Undersized targets (those less than 24 by 24 CSS pixels) are positioned so that if a 24 CSS pixel diameter circle is centered on the bounding box of each, the circles do not intersect another target or the circle for another undersized target; • Equivalent: The function can be achieved through a different control on the same page that meets this criterion; • Inline: The target is in a sentence, or its size is otherwise constrained by the line height of non-target text; • User agent control: The size of the target is determined by the user agent and is not modified by the author; • Essential: A particular presentation of the target is essential or is legally required for the information being conveyed.

or moving a finger or cursor along a specific path. This supports users who may find it difficult or impossible to perform gestures that require a significant degree of dexterity. This requirement does not prohibit the use of multipoint and path-based gestures where they would be appropriate, just that they should not be the only means of interaction. For example, an interface running on a touchscreen device might include zoom functionality that is supported by a pinching gesture, placing two fingers on the screen and moving them toward each other to zoom out and moving them away from each other to zoom in. Allowing zooming by double-tapping and providing zoom-in and zoom-out buttons are additional ways to support zooming without requiring a multipoint gesture.

Single-pointer actions involve a down-event—putting a finger or stylus on a control or pressing a mouse button while the pointer hovers on the control—and then an up-event—removing the finger, stylus, or releasing the mouse button while still on the control. Some users may inadvertently start a pointer action and need to cancel it without activating the control. When supporting single-pointer operation of a user interface element, design and implement the behavior so that the element's action associated with the pointer event is triggered only when the pointer event is complete. This allows users who start the pointer event to cancel the action by moving the pointer or finger away from the control before releasing. If this behavior isn't achievable, provide an easy way for users to stop or reverse the action.

Similarly, avoid relying on dragging actions, where a user must hold down a finger or mouse button while moving the finger or pointer from one location or another to access functionality. Providing a way for the same functionality to be operated by single-pointer actions, for example, by supporting clicking or tapping on the start location and end location, helps people who prefer to use pointer input but who find it difficult or impossible to perform dragging actions.

Some people may find it difficult or impossible to use functionality that is provided by motion actuation or sensor-based interaction, for example, where devices support input through shaking or tilting the device or using cameras for facial recognition to trigger an action. Where interaction is possible through motion actuation or sensors, design and implement additional ways to perform the interaction without requiring motion activation.

9.2.5.2 Supporting Speech Interaction   People who use speech input may announce the visible name of a control in order to place focus on the control for interaction. Speech input technology relies on matching the spoken input with the programmatic name of the control. If the programmatic name is significantly different from the visible name, the control may not respond to speech input of the visible name. This situation typically occurs when a control has been implemented in such a way that the accessibility information provided to assistive technology is different from the visible name, either by error or in an inadvertent attempt to address other accessibility requirements, in particular, for screen reader users.

For example, consider a button on a webpage with a visible text label, "Buy", and a different, accessible name, "Click this button to confirm purchase of the goods in your basket", provided using an `aria-label` attribute.

```
<button aria-label="Click this button to confirm
purchase of the goods in your basket">Buy</button>
```

This button can't be activated when a speech input user speaks the visible text ("Buy"). The accessible name is provided by the `aria-label` attribute rather than the visible text, and the value of the `aria-label` attribute doesn't contain the word "Buy". Removing the `aria-label` attribute ensures the control's visible label is also its accessible name. When a speech input user announces the visible name, "Buy", the speech input software will associate this name with the control, allowing speech activation of the control. Since the button is implemented using a `button` element and is announced by screen readers as a button, and the label communicates the button's purpose, there's no need to provide additional information explaining how to operate the button.

```
<button>Buy</button>
```

For any interface element that has a visible text label, make sure that the control's accessible name is the same as the visible name or contains the visible text.

9.2.5.3 Providing Usable Target Sizes The challenge of designing user interfaces so that it's easy to activate a control has been broadly recognized in general principles of usability and human computer interaction, where Fitts's Law[5] provides a formula to predict the time taken for a user to activate a control based on the distance between the pointer and the control and the size of the control. From an accessibility perspective, people with reduced dexterity and people with reduced visual acuity may require additional effort to move a pointer or finger to a specific control on the screen in order to activate the control. They may find it more difficult to activate the intended control and may inadvertently activate an adjacent control.

Make sure that the size of each control is sufficiently large and has sufficient space from adjacent controls to allow easy activation with reduced chance of error. WCAG provides minimum requirements for control size and spacing, but when designing the size, shape, and position of controls, pay particular attention to making it as easy as possible to activate each control without error.

9.3 Understandable

Information and the operation of user interface must be understandable.

This principle is broken down into guidelines with requirements for content to be presented such that it can be read and processed, that the interface's appearance and behavior are predictable, and that users are provided with necessary support when inputting data and dealing with errors.

Understandability requirements aim to address the accessibility needs of people with a range of cognitive disabilities that affect reading, learning, comprehension, memory, information processing, and the ability to recover from error states. They benefit people with anxiety, heightened stress levels, and mental health conditions that may make it more difficult to focus on content, which may increase the chances of making errors.

Meeting understandability requirements helps avoid introducing serious barriers for some people and also helps increase usability overall by improving content quality, consistency in behavior, and reducing the incidence and impact of errors.

9.3.1 Readable

Make text content readable and understandable.

Requirements under this guideline focus on specifying text content in a way that makes reading and comprehension easier for a digital product's target audience. Depending on the size and definition of "target audience," the ways to meet these requirements may be highly context-specific. Table 9.10 lists the constituent success criteria for this WCAG guideline.

9.3.1.1 Supporting Different Languages Accessibility requirements for language support focus on accurately identifying the content's language. This includes providing programmatic identification of the natural language of content and programmatic identification of language changes. These attributes support people using text-to-speech output by providing cues to screen-reading technologies to use the correct language and pronunciation. Language identification also supports automated translation between languages.

Engineering language support starts with establishing the default language of the product and identifying any instances where content or functionality is provided in a language other than the default. For example, in HTML, you can programmatically specify the language of a piece of content using the `lang` attribute with a value that corresponds to the language of the content.

For closed systems, support for different languages might require that you add functionality to change the natural language of the product, so the interface, content display, and speech output are available in multiple

Table 9.10 WCAG 2.2 Requirements for Guideline 3.1—Readable

Name and Conformance Level	Description
3.1.1 Language of Page (Level A)	The default human language of each web page can be programmatically determined.
3.1.2 Language of Parts (Level AA)	The human language of each passage or phrase in the content can be programmatically determined except for proper names, technical terms, words of indeterminate language, and words or phrases that have become part of the vernacular of the immediately surrounding text.

languages. For example, a kiosk might have functionality that enables users to choose a language at the start of an interaction. The setting then adapts the interface to use that language for all content and interaction.

Functionality to support readability could include options to switch between simple language and standard language versions of the same content, where available. Wikipedia is a good example of a digital resource that offers simple language versions of content.[6]

9.3.1.2 Supporting Different Literacy Levels Accessibility requirements for supporting a range of literacy levels focus on reading levels, with specifications related to providing content at a suitable reading level.

Engineering support for different literacy levels starts with using simple and clear language whenever and wherever possible. You can evaluate the reading level of content and aim for a reading level that is accessible to a broad audience. For example, WCAG SC 3.1.5 Reading Level specifies that content should be written in a reading level suitable for lower secondary education level.[7]

Designing for readability involves using typographic techniques to improve readability, such as breaking up text blocks into sections, with headings and subheadings providing visual and semantic signposts and bulleted lists to structure related groups of items. Rules and images are other visual ways of separating text into smaller chunks. Background color changes or borders can be used to highlight important text.

Look for ways to support text with images that help communicate the meaning of text. For example, include recognizable iconography along with text labels to help communicate the purpose of form and interface controls. When working with topic areas that are necessarily more difficult and complex, provide for alternative representations of the materials in your designs and layouts, such as making space for simplified summaries, illustrations, and graphs.

When producing written content, ensure the content meets readability requirements by following best practices for readability. This applies to all content, including instructions, notifications, dialog content, and error handling text.

- Write content in plain language to ensure an appropriate reading level for the target audience.
- Be careful when using idioms, metaphors, and other indirect ways of communicating meaning.
- Be judicious when using abbreviations. Where abbreviations are used, provide expansions on first use in a page or screen.

- Use specialist and technical words with care and define a glossary of terms and abbreviations that may be unfamiliar to some of the target audience.
- For long passages of content, prepare short summaries to provide at the start.[8]

A critical factor in supporting readability is implementing requirements for perceivability and operability. Specifically, implementing requirements that support flexibility in how text is visually presented means people can adapt content to meet their accessibility needs for reading. Also, implementing requirements that support compatibility and programmatic access allows users to read using assistive technology, such as reading assistance software, screen reader software, and braille output.

Ways to support readability through implementation include programmatically identifying the meaning of phrases that may not be familiar to the target audience, for example, by providing expansions for abbreviations and acronyms using markup or with custom tooltips that display definitions of abbreviations and unfamiliar terms. Implementation for readability in closed systems could involve providing an option to enable audio output for users with limited reading abilities, perhaps including highlighting features that visually show the content currently being read aloud.

9.3.2 Predictable

Make web pages appear and operate in predictable ways.

This guideline applies to consistency in the identification, location, and behavior of interactive controls. The requirement's intent is to reduce the cognitive load required to learn and understand how content is laid out and how functionality will behave across an interface. Table 9.11 lists the constituent success criteria for this WCAG guideline.

9.3.2.1 Enabling User Control   Accessibility requirements for supporting user control focus on enabling users to control their interactions with an interface using different interaction methods and on minimizing unexpected changes and interaction patterns. Requirements address how components should behave, for example, when they receive focus or accept user input. For example, users should be able to move focus to a menu and move through menu options without triggering a change of context. Examples of changes of context include loading a new page or screen, submitting data, opening a new browser tab or window, and moving focus to another element. Unexpected changes in context can be disorienting for many people, including people using screen readers or screen magnification and people with limited attention, increased anxiety or stress, or

Table 9.11 WCAG 2.2 Requirements for Guideline 3.2—Predictable

Name and Conformance Level	Description
3.2.1 On Focus (Level A)	When any user interface component receives focus, it does not initiate a change of context.
3.2.2 On Input (Level A)	Changing the setting of any user interface component does not automatically cause a change of context unless the user has been advised of the behavior before using the component.
3.2.3 Consistent Navigation (Level AA)	Navigational mechanisms that are repeated on multiple Web pages within a set of Web pages occur in the same relative order each time they are repeated, unless a change is initiated by the user.
3.2.4 Consistent Identification (Level AA)	Components that have the same functionality within a set of Web pages are identified consistently.

intellectual disabilities. Unexpected changes in context may not be noticed by some people, leading to confusion over the current state or location.

Designing interactions that support user control means avoiding introducing unexpected changes in context. For example, when specifying the behavior of form input elements, specify an explicit action a user must perform to submit the data, such as activating a "Submit" button. Another example is programmatically opening web links in a new tab or window, which is a change of context that is not initiated by the user. There may be situations that warrant opening content in a separate window or browser tab, but this behavior should be clearly communicated to users. When in doubt, honor a user's right to choose whether or not to open a link in a new window or tab.

9.3.2.2 Providing Consistency Accessibility requirements for consistency focus on making sure user interface elements are identified and presented in a consistent way throughout a digital product. For example, when a navigation mechanism, such as a skip link, menu, or list of navigation options, is presented across multiple pages or screens, the relative order in which the mechanism appears must be the same for each occurrence of the navigation mechanism. Interface elements that appear on multiple web pages and have the same functionality must be identified in a consistent way. This requirement applies to visible identification through text or imagery and to the text equivalent for imagery. These requirements help users locate, recognize, and understand the behaviors of navigation features that appear more than once, addressing a range of cognitive user accessibility needs as well as improving predictability and learnability for screen reader users.

Designing for consistency means striking a balance between applying a unique look-and-feel that distinguishes the interface from others and honoring design conventions that help reduce the potential for confusion. Consistency in design also means ensuring that multiple instances of elements intended to behave in the same way look similar and are specified with similar behavior on each page or screen where they occur. Consistency in visual presentation must be matched by consistency in text alternatives provided for each instance of buttons, links, and other interactive controls that are labeled by images, have the same functionality, and are repeated on multiple pages. Consistency in design also applies to the location of an element on a screen as well as its appearance. For example, for navigation menus that are presented across multiple screens or pages, the location and sequence of each menu should be consistent.

In development, consistency specified in designs must be honored when implemented in code, including ensuring that the visual layout of a screen or page is reflected in the code order. When implementing navigation mechanisms, make sure they are programmatically consistent in their location in the code order and sequencing from one page or screen to another. Pay close attention to elements like text alternatives for image-based controls and other non-displaying attributes that might not be visually apparent but where inconsistencies will affect assistive technology users. Meeting predictability requirements can be made easier by using coding practices that promote consistency of interface layout and behavior, including using templates and code libraries for user interface components.

9.3.3 Input Assistance

Help users avoid and correct mistakes.

Input assistance focuses on ways to reduce errors and where they occur to provide users with helpful error identification and recovery support. People with disabilities may be more likely to experience input errors and may require more time and effort to locate and recover from errors. Design that helps reduce the incidence of errors and provides sensitive, supportive, and low-effort error handling when they do occur particularly benefits people with accessibility needs while also improving usability overall.

In the context of this guideline, "input" covers a range of user interface elements that enable users to enter data. This includes entering text in a text input field or choosing an option from a predefined selection, such as a set of radio buttons, checkboxes, select menu, or date picker. Table 9.12 lists the constituent success criteria for this WCAG guideline.

Table 9.12 WCAG 2.2 Requirements for Input Assistance

Name and Conformance Level	Description
3.3.1 Error Identification (Level A)	If an input error is automatically detected, the item that is in error is identified and the error is described to the user in text.
3.3.2 Labels or Instructions (Level A)	Labels or instructions are provided when content requires user input.
3.3.3 Error Suggestion (Level AA)	If an input error is automatically detected and suggestions for correction are known, then the suggestions are provided to the user, unless it would jeopardize the security or purpose of the content.
3.3.4 Error Prevention (Legal, Financial, Data) (Level AA)	For web pages that cause legal commitments or financial transactions for the user to occur, that modify or delete user-controllable data in data storage systems, or that submit user test responses, at least one of the following is true: • **Reversible:** Submissions are reversible. • **Checked:** Data entered by the user is checked for input errors and the user is provided an opportunity to correct them. • **Confirmed:** A mechanism is available for reviewing, confirming, and correcting information before finalizing the submission.

9.3.3.1 Preventing Errors Accessibility requirements for preventing errors relate to providing helpful prompts and instructions that keep users on track when interacting with interface components, especially when providing and submitting data to the system. The objective is for users to successfully complete tasks and avoid making mistakes. The requirements focus on elements of the interface that can help users achieve this objective.

For user interface elements that collect input, a text label or instructions must be provided to explain the purpose of the element, including the data it collects and any formatting restrictions. This requirement helps ensure users know what information is needed and any formatting requirements before they provide the input, rather than after submission. The text label should be a succinct word or short phrase describing the element's purpose; instructions can provide more detail and should be placed sufficiently close to the element they describe. Examples of essential instructions might be a visible indication that a field is required, requirements that must be met for a new password, or the required format for a phone number or date. In certain specific situations, a recognizable icon might be used as a label for a control, such as a magnifying glass for a search input. If icons are used as labels, they must be provided with an equivalent text alternative. Providing a visible text label to supplement the icon is preferable, and if this is not possible, ensuring the text label appears as a tooltip on mouse hover or focus helps reinforce the icon's meaning for people who are unfamiliar with the icon.

In some cases, providing clear labels and instructions may still be insufficient to help users understand how to use specific functionality. That means a way should be provided for users to access context-sensitive help where input is required. This help could be provided in custom content and, depending on the platform and technology used, may also be automatically available through programmatically available information about an input element. Context-sensitive help extends requirements for labels and instructions to cover additional information that may help users in general avoid making errors and, in particular, support people with a range of cognitive user accessibility needs in understanding what input is needed and how to provide it.

For data input events that cause financial transactions or legally significant actions, submit test responses, or edit or delete user-controllable data, a way must be provided to either review and correct the data before submission or reverse the submission after it has happened. For example, the submission process could include an intermediate stage where a user's input is presented for each input field and options are provided to return and edit input or confirm submission. This requirement addresses specific situations where an input error can have particularly significant and adverse consequences for users, such as an inadvertent purchase, legally binding decision, or deleting critical data. Support for reversing input decisions could also be extended to other situations where input errors have less significant consequences. Reassurance of a way to reverse significant actions can help people with heightened anxiety and stress levels, while also helping to reduce the potential consequences of unplanned spending or other legally significant decisions—a behavior associated with certain mental health conditions such as bipolar disorder or depression.

When designing user interactions, pay close attention to what's needed to reduce the likelihood that users will make mistakes or slips.

- Provide labels and instructions that clearly communicate the purpose of input elements and provide sufficient information to allow users to interact with each element, including details of any restrictions on input, such as specific formatting requirements or ranges of accepted values.

- Identify input elements and other functionality that may be unfamiliar to some users and that require additional help. Provide easy access to clear and understandable, context-sensitive help.
- Design ways that users can review and edit input choices before submission, particularly for legally and financially significant input actions and where significant changes to user-editable data will occur. For example, some financial organizations have taken approaches to address potentially unintended purchases by delaying processing transactions made late at night, alerting users the next day, and offering an option to revoke the transactions.

Correct implementation of error prevention features is important so that all users benefit from the support. Ensure all input elements are provided with programmatically defined labels and that any instructions or input requirements, such as required status, are programmatically defined and available to assistive technology.

9.3.3.2 Providing Error Notifications Accessibility requirements for error notification focus on making sure systems provide feedback when users do make slips or mistakes to help them recover from the error and get back on track to meeting their objectives.

When an input error made by a user is automatically identified, the error must be described in text with sufficient detail to allow the user to understand the nature of the error and how it may be resolved. Examples of input errors that can be automatically detected include when a user:

- Does not input data in a field that is required.
- Provides an input value that is outside of an accepted range or set of values, such as a non-existent zip code or postal code, a number that is higher than a maximum quantity allowed, or an end date that is earlier than a previously specified start date.
- Specifies an input value that does not meet the required format, such as a date or phone number.

Some input errors may not be automatically detectable, so this requirement only applies to those errors that can be detected programmatically. In some cases where errors could be detected, error identification is not required if it would affect the purpose of the digital product or cause security issues. For example, in a form used for online assessment in an educational context, it might not be appropriate to identify and inform users of incorrect answers to test questions while they're still taking the test.

For this requirement, it's not sufficient to identify the presence of an error through limited visual means, for example, redisplaying the input fields after a user attempts to submit the form or by highlighting a field with an error in a different color. Visual identification can certainly help with identifying the presence and location of errors, but the text description is essential for accessibility. The text description must be presented in plain language, so providing an error code or an overly technical description of the error will not meet this requirement.

For input errors that have been automatically identified and where guidance for recovering from the error is known, this guidance must be provided to users. Guidance could take the form of a short text string explaining an action a user could take, such as suggesting a list of acceptable values. Programmatic indication of the error in a way that draws attention to the error, for example, through a dialog, can help ensure that users are aware of the error suggestion. Although not a requirement for error suggestion, providing a notification in text of *successful* input submission can provide reassurance to users that data was successfully input.

When designing and implementing error notifications, focus on approaches that make the notification clear, helpful, and accessible to all users.

- Design error notifications for easy identification of the error notification and the source of the error.
- Write error messages that clearly describe the nature of the error and provide users with sufficient information to help them recover from the error.
- For error-checking functionality, be as liberal as possible with input within the specified restrictions of the application and platform. For example, scripts that accept different forms of phone numbers, including plus, bracket, and dash characters that some people might add when writing a phone number, and then strip out those characters, are more supportive of diverse input than scripts that reject any non-numerical input.

As with most controls and interactive elements, ensure all error messages are included in efforts to meet requirements for programmatically identifying relationships between content. Implementation techniques that help to meet input assistance and other understandability requirements overlap with the final accessibility principle, *Robust*, which we cover in the next section.

9.4 Robust

Content must be robust enough that it can be interpreted by a wide variety of user agents, including assistive technologies.

This broad accessibility principle is represented by a single guideline that addresses compatibility of content and functionality with the diverse range of technologies that people with accessibility needs may be using to access digital products, now and into the future. It collects together key accessibility requirements that are not already covered under Perceivable, Operable, and Understandable principles.

Requirements for robustness address the accessibility needs of people who use current assistive technologies that present digital content in different modalities, such as screen readers, addressing a range of user accessibility needs. These requirements are also forward-looking, anticipating the needs of assistive technology yet to be developed.

9.4.1 Compatible

Maximize compatibility with current and future user agents, including assistive technologies.

The requirements for compatibility address different ways that content and functionality can be coded to work with a range of technologies, including assistive technologies, and encourage an implementation approach aimed at future-proofing accessibility in digital products. This guideline specifically mentions user agents, indicating its focus on web content. But the principle of ensuring compatibility with assistive technologies encompasses all digital interfaces and includes open and closed technologies.

In this section, we concentrate on two requirements of this guideline— name, role, and value accessibility information, and status updates. A third WCAG 2.1 requirement addresses specific parsing requirements for resources written in markup languages such as HTML. With the evolution of browsers and assistive technology, this requirement has been judged to be sufficiently less relevant to ensuring accessibility of web content and doesn't appear in WCAG after version 2.1. Table 9.13 lists the constituent success criteria for this WCAG guideline.

Table 9.13 WCAG 2.2 Requirements for Guideline 4.1—Compatible

Name and Conformance Level	Description
4.1.2 Name, Role, Value (Level A)	For all user interface components (including but not limited to: form elements, links and components generated by scripts), the name and role can be programmatically determined; states, properties, and values that can be set by the user can be programmatically set; and notification of changes to these items is available to user agents, including assistive technologies.
4.1.3 Status Messages (Level AA)	In content implemented using markup languages, status messages can be programmatically determined through role or properties such that they can be presented to the user by assistive technologies without receiving focus.

9.4.1.1 Supporting Programmatic Access to Interactive Controls Accessibility requirements for interaction focus on ensuring that each interactive element in an interface is implemented with core accessibility information that can be utilized by assistive technology. Exposing an element's name, role, and value or state information allows assistive technology to provide these details to users so they can decide whether and how to interact with the element. This requirement extends to ensuring that changes to state or value are also updated in an accessible way so they can be communicated to users.

This requirement places responsibilities on the design and implementation of user interface elements to ensure that critical accessibility information is provided. These responsibilities vary in significance depending on the platform being used to build the digital product, the complexity of the user interface, and the approach taken to construct the user interface.

At the design stage, responsibility for meeting robustness requirements starts by choosing the most appropriate user interface element for a specific action. Matching an interface element for the action it's intended to support reduces the work required to provide it with appropriate accessibility information. By contrast, choosing an interface element and then adapting it for a different purpose creates extra implementation work in changing role and value information for the new purpose of the element. Usability issues can result when an interface element's actual behavior does not match its intended behavior.

A second responsibility at the design stage is specifying name, role, state, and value information (or ranges of possible states or values) for each user interface element in the design specifications passed to development for implementation. Ensuring this is part of the design process helps to ensure that developers have what they need to implement the user interface element in an accessible way, rather than guessing what information might be needed.

When implementing user interface elements, using platform native interface elements will typically ensure that name, role, state, and value information is accurately provided by default to assistive technology. Where a design specification includes custom interactive elements not provided natively by the development platform, additional development work is needed to provide each element with the necessary accessibility information and to ensure it remains accurate as the element changes state or value.

For web applications with complex user interface elements, native HTML may not provide sufficient capability to provide the desired functionality in an accessible way or may not yet be sufficiently supported across target browsers. The WAI-ARIA specification introduced in Chapter 6, *Guiding Principles*, provides a large set of supplementary HTML attributes and attribute values that can be used to give custom user interface elements the necessary name, role, value, and state information. To help developers of complex web applications, the ARIA Practices Guide (APG) offers a definition of how ARIA can be used for a wide range of user interface patterns, including tab panels, carousels, accordions, dialogs, editable grids, feeds, and toast notifications.

Support for ARIA across browsers and assistive technologies is strong and improving, but at the time of writing, it is not complete. This means it's important to understand the potential limitations of ARIA and use ARIA only with extreme care when building web applications. Incorrectly implemented ARIA can reduce rather than improve accessibility. ARIA attributes override native HTML elements and attributes when exposing accessibility information to assistive technology. If misused, it can replace correct accessibility information about an element with incorrect information, such as the wrong role or an inaccurate name. Relying on ARIA also increases overhead for maintaining code and for tasks such as internationalization and localization. Wherever possible, start with native controls and use ARIA only when there is no other way to achieve the desired user interface.

Every engineer should know... custom UI components need extra information to be accessible.

By Kate Kalcevich

Interactions on the web are notorious for being inaccessible to people with disabilities and are often part of the most critical functions, such as:

- submitting a job application,
- purchasing from an online store, or
- booking a healthcare appointment.

When you use HTML form elements like `<input>`, `<select>`, and `<button>`, information about the element is passed to the DOM (Document Object Model) and into an Accessibility Tree. Assistive technologies that disabled people use to interact with websites and apps can access the nodes of the accessibility tree to understand:

- what kind of element it is by checking its role, e.g., checkbox;
- what state the element is in, e.g., checked/not checked;
- the name of the element, e.g., "Sign up for our newsletter."

The other thing you get with HTML elements is keyboard interactivity. For example, you can access a checkbox using the TAB key and select it using the SPACEBAR.

When you build your own custom components or use ones from a framework, you need to provide information about the element using ARIA and build keyboard interactivity for assistive technology users. Static elements like <div> and <span> that are commonly used to create components from scratch have no role, state, or keyboard access.

Accessible Rich Internet Applications (ARIA) is a technical specification that includes roles and attributes that help make web content and applications more accessible.

A role tells an assistive technology user what an element is. A button is very different from a banner. Choose a role that matches the function of the component you're building.

ARIA roles override an HTML element's inherent role. <img role="button"> is no longer an image but a button. Use with extreme caution.

If an element has a state (e.g., hidden, disabled, invalid, readonly, selected, and so on) or changes state (e.g., checked/not checked, open/closed, and so on), you need to tell assistive technology users what its current state is and its new state whenever it changes.

While ARIA attributes tell a user what the state is, you still have to write code to make it true. aria-checked="true" doesn't actually check a checkbox; it just tells the user the checkbox is checked.

Anything interactive on a website or application must be able to receive focus. Every element you can click on should also be accessible using the keyboard.

There are three concepts to remember if you use static elements like <div> and <span> to build interactive components:

- You need to add tabindex="0" so that a keyboard or emulator can focus on them.
- For anything that accepts keyboard input, you need to add an event listener to listen for key presses.
- You need to add the appropriate role so that a screen reader user can identify what element you've built.

When you're building a custom component, check the W3C ARIA Authoring Practices Guide to figure out what ARIA attributes and keyboard interactions to add.

You can validate the ARIA you've written by using:

- Accessibility browser extensions like axe or WAVE
- Accessibility linters like axe for Visual Studio Code or ESLint for JSX elements
- A screen reader to listen to the component
- Assistive technology users to test it

Testing with users with disabilities should be a standard practice in any large organization. The users we build digital products for are the experts in what works for them and what doesn't when it comes to accessibility.

A longer version of this sidebar appeared in *Smashing Magazine*: https://www.smashingmagazine.com/2022/09/wai-aria-guide/

9.4.1.2 Providing Accessible Status Updates When status updates are presented, they must be presented in a way that programmatically identifies them as status updates, so that assistive technologies can provide the update without having to move focus to the status update. This ensures that assistive technology users who might not otherwise be aware of the status update are informed of the update without being disrupted from their current point of focus on the page or screen.

For status notifications, using native methods for updates can help ensure they are automatically exposed to assistive technology with a level of disruption appropriate to the significance of the notification. For websites and applications, ARIA extends HTML to provide live regions, which is a way to identify a container (or potential container) of notification content. When content in a live region is updated, the update is detected and reported by assistive technology. For example, if a script populates a container with a notification message, a screen reader will detect and announce the message. Live regions have a value of "politeness" specified, which governs whether the screen reader announces the region update after reading out the currently focused content or interrupts current content to announce the update, which may be necessary for more significant notifications.

Takeaways

As an engineer, you should:

- Understand and be able to articulate the scope and intent of the Web Content Accessibility Guidelines, and apply that understanding to other standards that incorporate WCAG requirements.
- Apply the intent of accessibility requirements to a range of technologies, platforms, and digital interfaces.
- Realize the intricacies of accessibility requirements and that engineering digital accessibility is not a straightforward checkbox-type activity.
- Be ready to delve into and become fluent in accessibility requirements; develop mastery over requirements that fall within your domain of responsibility in engineering digital products.
- Approach accessibility and meeting accessibility requirements as a core professional practice and responsibility for all product development.

Notes

1 L. Watson (2011) *Text descriptions and emotion-rich images.* tink.uk/text-descriptions-emotion-rich-images

2 R. Whitaker (2020) *Developing Inclusive Mobile Apps—Building Accessible Apps for iOS and Android.* New York: Apress.
3 W3C (2022) *Page Structure.* www.w3.org/WAI/tutorials/page-structure/regions
4 W3C (2018) *Input Purposes for User Interface Controls.* www.w3.org/TR/WCAG21/#input-purposes
5 *Fitts's Law.* en.wikipedia.org/wiki/Fitts%27s_law
6 *Wikipedia.* www.wikipedia.org
7 W3C (2023) *Understanding SC 3.1.5: Reading Level (Level AAA).* www.w3.org/WAI/WCAG21/Understanding/reading-level
8 PLAIN. *What is plain language?* plainlanguagenetwork.org/plain-language/what-is-plain-language

10

DESIGN AND DEVELOPMENT

Objectives

Our objectives for this chapter are to explore ways to meet accessibility requirements during the design and development phases of product development. Starting with the concept of inclusive design, we explore how design and development approaches can be oriented toward meeting accessibility requirements in the most successful way. We cover ways to integrate accessibility into design and development practices in order to promote, support, and sustain attention to accessibility throughout the product lifecycle.

Once you're through this chapter, you should:

- Appreciate how attention to inclusive design in design and development approaches can support accessible outcomes.
- Understand ways requirements can best be approached and managed at different phases of product development.
- Understand ways to adapt design and development practices to effectively support and manage accessibility.

Introduction

The Inclusive Design Research Center at the University of Toronto defines inclusive design as "design that considers the full range of human diversity with respect to ability, language, culture, gender, age, and other forms of human difference."[1] In other words, inclusive design is a process that pays deliberate attention throughout the lifecycle of creating a digital product or resource to understanding and accommodating people with diverse needs. More specifically, inclusive design is an intentional approach to ensuring

DOI: 10.1201/9781003288060-12

the needs of under-represented and historically excluded user groups are met. Inclusive design provides a helpful framework for considering how best to integrate accessibility requirements into design and development approaches and practices.

In Chapter 5, *Core Attributes*, we introduced a set of core accessibility attributes, and in Chapter 9, *Core Requirements*, we explored a core set of accessibility requirements that apply across a wide range of digital products. In addition to these core requirements, for any given digital product, the specific objectives and context of use will present additional requirements that need to be met to ensure that people with disabilities can successfully use the product. Here we cover design and development approaches that best support implementation of accessibility requirements and design specifications and show how recognized approaches can be aligned with meeting accessibility needs. Within each approach, we discuss ways to integrate accessibility considerations into development activities and support continuous attention and improvement, with the aim of embedding accessibility and inclusive design into standard practices.

10.1 Design Approaches

Attention to accessibility in the design phase is critical for ensuring accessibility requirements are met in information, visual, and interaction design. Decisions that affect color, typography, layout, and use of images and iconography can affect accessibility, particularly for people with vision impairment or with reading, comprehension, or attention difficulties. Design choices for user interface patterns for specific functionality also impact accessibility, given the potential impact on cognitive load for users and the effort required to operate functionality. Design choices also influence the burden on development when implementing patterns in the most accessible way.

When design decisions are made that create accessibility barriers, the chances that these barriers can be overcome in the implementation stage are low. For example, a design decision to use a low-contrast color palette may be difficult to reverse at the development stage. Accessibility details for a custom component and interaction patterns must be specified in the design so that development knows how to implement the component to support accessibility.

On the other hand, design can achieve greater accessibility when accessibility is a design requirement. Design approaches and specifications can explicitly address accessibility needs, making it easier for development to implement accessible products. The Inclusive Design Principles[2] can help with the process of design decision-making with inclusion in mind.

10.1.1 Provide Visual and Structural Semantics

Information design communicates meaning, or semantics, through visual attributes and positioning. In a book, for example, a section heading may be designed to be larger than the surrounding text and have a heavier weight, like bold or semi-bold. It may be a different typeface and color. Often, a heading is more proximate to the following text than the preceding, with more spacing before than after. Additionally, headings convey the information structure of a document. The main heading is larger than section headings, which in turn are larger than subheadings. These visual cues help users serve as signposts, helping readers navigate through the document, find relevant information, and improve understanding.

With accessibility, the job of a designer in this context does not stop with designing visual characteristics, such as the font, styling, and before and after spacing of a subheading. The designer must also design the semantics of information and interface design as part of good design practice. In the design phase, it's important to design to convey both the visual semantics and to annotate the programmatic semantics so that developers can implement the elements using the correct semantic markup in the development phase.

For accessibility, attention to visual semantics is important to clearly communicate information structure and the relatedness of elements. For example, since people using screen magnification may only see a segment of a screen at a time, proximity and grouping become key to communicating how elements are related. Clearly defined and delineated information hierarchies, signposted by clear headings, are easier to decipher and navigate, and reduce cognitive load. Additionally, as we've learned, content semantics can be encoded so that the relationships and hierarchies conveyed visually can be conveyed and acted on by assistive technology. For example, attention to the proximity of related information helps ensure that important related content such as a form field and its label remain visually close when a screen is magnified. The important point here is that information design also includes designing how things will be implemented in code so that information structures can be communicated in multiple modalities, including using text-to-speech tools like reading assistance and screen reader software, and when visual appearance is transformed to meet specific user needs.

Screen and page layouts benefit from clear and consistently defined structural semantics, with defined regions that allow users to quickly scan and recognize different functional areas of a design and move their focus to the most relevant region for the task at hand. For instance, most software has a menu region at the top of the screen where users know to look when they need to accomplish tasks. These regions often contain sub-elements; for example, a website header often contains a search feature, and users know to look to that site header region when they need search functionality.

For accessibility, consistent and clearly delineated regions help users understand the information structure of a screen or view. Visually enclosing regions can help users keep track of where they are in a page or view. For example, a uniform background color in a site menu region or section navigation menu helps screen magnification users stay oriented to their location in a magnified view. The background color marks the region and defines the enclosed elements as connected in the same region. Like content semantics, page structure can also be programmatically defined so that people who do not have access to the visual signifiers can understand, scan, and navigate readily to the most relevant region. Screen reader users, for example, can navigate the regions of a screen when those regions have programmatic markup. With encoded page structure, software can provide functionality to support scanning the page, moving between regions, and skipping over irrelevant regions.

10.1.2 Support Adaptation and Progressive Enhancement

The core attribute of Adaptability that we introduced in Chapter 5, *Core Attributes*, underpins the need to make design for digital products a process of creating designs and layouts that can then be adapted by users to fit their needs and preferences. This attribute makes the process of digital design fundamentally different from many other examples of product design, where adaptability options are more limited. In the case of digital design, the designer's role is to provide a design that fits the primary use case and meets baseline requirements, but to do so in a way that accounts for variation and adaptation. This approach is sometimes referred to as progressive enhancement, where the design starts from a baseline of design and usability that can be progressively improved depending on the context of use.

Where the platform provides flexibility, the visual design process should extend to catering for different zoom levels, text sizes, and screen widths. A common example of progressive enhancement is responsive web design. With responsive design, a website or application is optimized to adapt the user experience across different devices and viewports. When viewed on a small smartphone screen or large monitor, page elements change position or are suppressed at smaller screen widths. The role of design is to ensure designs are sufficiently flexible to support different screen widths, which also helps accommodate people who zoom in or increase text to a larger size.

One key aspect of progressive enhancement is the separation of content and presentation, as we discussed in Chapter 5, *Core Attributes*. Being able to apply different formatting to content and interactivity makes it possible for users to apply adaptations that meet accessibility needs. For example, high-contrast mode supports users who need a dark background and light text to distinguish text and interface elements. When designing digital products, keep what you are working with from a content perspective separate from how the content will look. Design for different views, such as different window widths, color settings, and text sizes.

10.1.3 Design for Multiple Use Cases

When we think about inclusive design and designing for the broad range of human needs and preferences, there is often no single approach that meets all user accessibility needs. In some cases, the most accessible design is a redundant design, where different content and attributes that serve the same purpose are designed into the interface. Using equivalent methods to achieve the same purpose helps support multiple and diverse use cases.

For example, color is a valuable design tool for distinguishing elements, conveying relationships, grouping elements, and conveying style and identity. Careful use of color to communicate information and relationships and to convey a suitable aesthetic value can significantly enhance usability and the user experience. However, some users may not perceive the colors as intended, including screen reader users, people with color deficits, and people who change how color displays. This means color isn't a reliable tool when used on its own. For example, when using a red border around input fields that have input errors, the issue isn't the use of a red border but rather *only* using a red border. The solution is to provide the information in multiple modalities. As well as showing sold-out items in red text, include an icon (with a suitable text alternative) or text label that provides the same information.

Another example is interaction feedback, providing affordances and feedback to help users operate an interface. For example, operable controls in an interface should change their visual appearance when they are focused, to communicate what control will be activated if selected. This visual indication is often triggered by the user hovering over a control using a pointing device, such as a mouse. But some users don't operate interfaces using pointing devices and therefore won't receive this feedback and won't know where they are in an interface. Therefore, interaction design for accessibility must take into account different ways of interacting with a product, for example, by providing a clear and consistent focus indicator to support sighted keyboard users.

Supporting universal access means approaching design from multiple directions and being ready to apply a range of design approaches to address different use cases and user accessibility needs. One helpful approach to keeping diverse user needs in mind throughout the project is to create a set of personas representing people with accessibility needs, ideally based on data gathered from your own up-front user research. A useful sample set of personas with accessibility needs is available in the book, *A Web for Everyone*.[3] Personas as representations of characteristics, needs, and motivations of target audience members can then be referenced during design decision-making, helping you to answer questions such as:

- What do we need to do to make sure that Persona X's user needs are met?
- If we make this decision, what impact will it have on Persona X?

10.1.4 Favor Simple and Clear Designs

Graphic design creates a visual logic on pages and screens through the judicious use of visual emphasis and variation. Edward Tufte, a pioneer in information design and data visualization, explains graphic and information design as "differences that make a difference" and notes that "the most economical means can yield distinctions that make a difference."[4] With this in mind, visual logic is best communicated on a backdrop of simplicity, where important features can stand out, as opposed to complexity, where everything is emphasized and nothing stands out. Minimizing unnecessary clutter and distractions allows important elements to stand out. Clear and simple graphic design guides users through information and functionality following a logical path and flow, engaging users visually while also conveying information about content and functionality.

Overly complex designs can cause accessibility barriers—even seemingly minor issues of inefficiency or redundancy can have a substantial cumulative impact for users who have to use a system regularly or repeatedly. Complex layouts with multiple columns and sidebars and complex tables with nested header rows and expandable content make it more challenging to meet requirements for meaningful sequence and focus order. In turn, this places additional overhead on development and testing to implement designs in the most accessible way. A design focus on minimizing unnecessary complexity helps reduce overhead while also encouraging creative ways to design complex content and functionality in alternative ways.

10.1.5 Use Established Design Patterns

Interaction design involves conveying information about content and functionality through the artful use of visual characteristics to guide the user through information and show them how to work with functionality. Semantics in the context of interface design is about assigning semantic meaning to content and functionality—a heading, a button control, an input label, a tab panel component. The best approach to accessible interaction is to use native controls that are fit for purpose and benefit from their default affordances, both visually and semantically. For example, a button has built-in visual characteristics that tell users how it works and what to expect. A programmatic button element provides the same affordances for nonvisual users. When elements with established affordances are used incorrectly, accessibility is impacted, for example, when table elements are used to create column layouts. Accessibility is also impacted when elements that have established affordances are not used, for example, when non-semantic <div> elements are used to create buttons rather than a <button> element or when bold and large text is used to provide a section heading rather than a heading (for example, <h1>) element.

Development has a significant role to play in coding accessible user interface widgets, but design plays an equally significant role in choosing the most appropriate user interface element for the interaction it's intended to support. Take advantage of conventions that users have learned and remembered, such as linking the website's logo to the homepage and including a "Contact us" link in the website footer. These pattern affordances can improve usability and accessibility when used consistently and according to convention. They can have a significant negative impact when applied incorrectly, going against user expectations and forcing users to unlearn and relearn interactions. By adopting the pattern affordances as design patterns within design systems and applying them consistently, you gain usability benefits through internal consistency with your products and within the digital landscape more broadly.

Every engineer should know... accessibility requirements for neurodivergent people.

By Lē Silveus

While adhering to Web Content Accessibility Guidelines (WCAG) is a significant step, it's not the whole picture. Neurodivergent individuals often face unique challenges when interacting with websites and applications. Some common issues include:

- **Sensory Overload:** Complex designs, rapid animations, high saturation colors, high contrast, and overwhelming visual or auditory stimuli can cause sensory overload for neurodivergent users.
- **Cognitive Load:** Complicated navigation structures, unclear instructions, and dense content can increase cognitive load and hinder effective interaction.
- **Attention Difficulties:** Neurodivergent individuals might struggle with focusing on relevant content amid distractions or overly busy layouts, or they could be so pulled in by attention-stealing tactics that harm is done in their day-to-day lives.
- **Input Sensitivities:** Some users might have difficulties with precise mouse movements, complex keyboard shortcuts, or touch gestures.

Next steps for advancing support for neurodivergent people include:

- **Building Diverse Teams:** Diverse teams bring diverse perspectives. Including neurominority individuals in your development process ensures that a broader range of user needs and experiences are considered. This can lead to innovative solutions that benefit everyone.
- **Elevating Neurodiversity in Accessibility Considerations:** When thinking about accessibility, go beyond the standard checklist. Consider the specific challenges faced by neurodivergent users. For instance, opt for clear, straightforward language, minimize distractions, and provide customizable settings to cater to different sensory preferences.
- **Harnessing Usability Testing with Neurodivergent Users:** Usability testing involving neurominority users can reveal insights that might otherwise be overlooked. Their feedback can uncover usability issues that, once resolved, lead to a more seamless and enjoyable user experience for all.

The principles of simplicity, clarity, and customization emerge as key strategies to enhance usability and provide an inclusive online environment. As we move forward in shaping the digital future, let's ensure that no user is left behind.

10.1.6 Annotate Accessibility in Design Documents

A critical responsibility of design is to provide sufficient information to developers to allow designs to be implemented with accessibility in mind. The alternative is to hand over a design without accessibility documentation, leaving development teams to guess how best to provide accurate accessibility information as they code the design, increasing the chance that accessibility is inaccurately coded or forgotten about. Annotation is the bridge between design and development. By supplementing visual designs with written information in the form of design annotations, you can communicate a range of accessibility requirements. Adding annotations directly to designs ensures that all information is together in one place, though you may also wish to generate a document with annotations as a backup.

You can communicate several types of essential accessibility information through design annotations. These annotations cover different dimensions of design, including visual design, user experience and interaction, and content design. Annotations typically accompany design artifacts, such as high-fidelity components and screen mockups. The following are accessibility features that can be communicated through annotations as part of the design specification.

- **Semantic Information for Content:** For designs for the web and other platforms that support semantic markup, annotations can communicate a content item's semantics. This might include section headings, lists and list items, table captions, row and column headers, and blockquotes. Annotation can also be used to communicate information about regions of a page, for example, to indicate the extent of the header, footer, and navigation regions. All of this information can be used by development to ensure structure is accurately represented in markup.

- **Focus Order:** Notate a logical and predictable focus order through interactive elements. Ideally, focus order should be logical, following the visual order of elements—for left-to-right languages like English, that would be starting top left and finishing bottom right of the screen. But for multi-column layouts and complex user interface components where order may not be clear, the risk is increased that implementation creates an incorrect focus order. Annotation indicating numbered focus order helps unambiguously communicate the order in which active elements should receive focus. Annotation can also emphasize how focus should be managed when content is added or removed from a screen in specific states.

- **Reading Sequence:** Related to focus order, annotations can emphasize in which order content should be read out by a screen reader. Again, this is most helpful for visually complex layouts, such as multi-column layouts, where, without clear guidance, the intended reading order could be affected by coding decisions.

- **Components and Design Patterns:** Specify which elements (ideally native elements) to use for interactive elements and components. When an appropriate native element is not available on the platform, include notations for custom components specifying accessibility features, including name, role, and state information and keyboard interaction.

- **Accessibility Information:** For user interface components, annotations should be used to clearly indicate what kind of component it is (for example, a button, checkbox, or radio button), what its accessible name is, and what state or value it holds at specific instances of the design. The same information should be provided for user interface components that are collections of components, like tab panels, carousels, and disclosure widgets. This is essential information for development to use when implementing the user interface in an accurate way.

- **User Interface States**: Provide and annotate accessible designs for different states for user interface (UI) components, such as hover and focus states for buttons and links. Annotate accessible designs for different states, for example, "you are here" indicators showing which item is selected in a tab panel or menu, or error notification indicators for inputs in error.

- **Labels and Names:** Provide control labels, including text labels for image-based controls like icons. Flag which images are decorative and which are informative. For informative images, include text descriptions along with the design annotations so that developers can easily add the text into code.
- **Alternative Text for Images:** Where designs include images, such as icons as labels for controls, annotations can be used to communicate the text alternative for the image. An annotation can also communicate that an image should have an empty text alternative, indicating it should not be presented to assistive technologies.
- **Color Values:** Designs and supporting documentation should communicate accurate color information for development to use when applying visual styling to the user interface. Assuming color schemes have been chosen with minimum contrast ratios in mind, this helps ensure that development honors specified colors when implementing designs.

10.2 Development Approaches

The development phase is where accessibility requirements are implemented in code. In Chapter 9, *Core Requirements,* we covered a significant number of requirements that relate specifically to how content and functionality are provided. Many of the core accessibility requirements are addressed in development rather than design or content. This means attention to accessibility in the development phase is essential to meeting accessibility requirements. It's safe to say that digital accessibility cannot be successfully achieved without close attention to accessibility in development.

In addition to implementing core accessibility requirements, development must also implement accessibility decisions made at the visual and interaction design stages. Working from design specifications and annotations, development must make decisions about how best to implement accessibility and seek guidance from other roles in the product team before making coding decisions that affect accessibility and user experience.

10.2.1 Use Platform-Native Controls and Components

Development approaches to implementing controls and components have a major impact on accessibility. Controls are single-purpose interface elements, like menus, buttons, and links. Components, or widgets, are collections of controls that together form a more complex user interface element, such as a sortable data table, date picker, tab panel, or carousel. User interface controls and components typically require meeting a number of related requirements, which can be summarized as a few related development goals:

- Each control is provided with accessible name, role, state, and/or value information. This information is updated to reflect any change in state or value.
- Each widget, or collection of controls, also has an accessible name and role and is programmatically identified as a related group of interdependent elements.
- Each control is focusable and operable using the keyboard and appropriate swipe gestures for touchscreen-based interactions. Platform conventions for operating the control are supported.
- Each control has a clear visual indication when it receives keyboard focus.
- Any related information for the control, such as an error message or instructions on how to operate the control, is programmatically associated with the control.
- Any error message or other notification of a status update is communicated in an accessible way.

Implementation approaches for meeting the requirements depend on the platform. In some cases, the platform will come with elements that match the design requirements. Many modern technology platforms have embedded accessibility into native platform elements, which means using platform controls and widgets can help reduce the effort of meeting development goals. Whenever possible, use platform-native controls and components.

In some cases, a design may specify content and functionality that do not have a clear match with platform options. The decision to create a custom control or component should be carefully measured and discussed. First, work with the design team to determine whether there are other design approaches that would provide the needed content and functionality and that can be implemented using a platform element. If the required design cannot be achieved using platform-native controls and components and a custom implementation is required, you'll need to provide the necessary accessibility information and interaction to meet the development goals of controls and components.

10.2.2 *Customize Accessibility When Required*

Implementation of accessible controls and components can benefit from augmenting accessibility features in native platform elements or creating custom implementations for content and functionality that's not achievable using platform options. Some platforms provide options for implementing greater accessibility than might otherwise be provided by out-of-the-box controls and components.

For example, with web development, ARIA makes it possible to create custom-accessible interactive elements. However, using ARIA comes with overhead and unpredictability that is often unnecessary. As an example of how ARIA can be used to augment native HTML to provide critical accessibility information about a user interface element, let's consider an example of a custom button. Some web development approaches involve custom buttons implemented using `<span>` elements and labeled using CSS icons, with behaviors provided using JavaScript.

This is a code example of an HTML custom button implemented with a `<span>` element and using a CSS-supplied icon as a button label:

```
<span href="…" class="submit"></span>
```

Without additional accessibility efforts, this implementation can cause problems. Assistive technology (AT) will not announce the element as a button because it has no accessible role. Using AT features to locate buttons does not work for this button since it doesn't have a button role. Since the button is a `<span>` element, which is not interactive, it is not included in the focus order and can't be focused by the keyboard. Even if scripts providing button behaviors are written to include keyboard operation, the button isn't operable using standard keystrokes.

It's possible to add code to ensure the button has an accessible role and to include it in focus order so that keyboard operation is possible. Here, the code includes the `tabindex` attribute, which makes it operable, and the `role` attribute so that it's announced as a button.

```
<span role="button" tabindex="0"
aria-label="submit"> </span>
```

This approach requires extra code, which must be maintained.

The simplest and most robust approach is to use a native button element, which by default includes the correct role and is automatically included in focus order, so keyboard users can move focus to the element in order to interact with it. Here is the custom button implemented using a `<button>` element:

```
<button type="submit">Submit</button>
```

In this simple example of a button, it may not be worth investing the time and attention to create a custom "submit" button for a web form interface. HTML provides native elements that perform that function, including `<input>` and `<button>`, so there's unlikely to be a strong case for implementing a custom button in a website or web application.

In some cases, extending and customizing controls and components to meet accessibility requirements is worth the investment and may be unavoidable, particularly when working with platforms and technologies that are less advanced than the web in native platform accessibility support.

The important consideration in the development phase is to avoid heading down the path toward customization unless it's completely necessary. When you do extend and customize code for accessibility, treat those customizations as essential features that you maintain throughout the product lifecycle.

10.2.3 Check for Accessibility during Development

When designs are provided with sufficient annotations communicating accessibility information for a user interface and its constituent elements, use this information to guide implementation. Where information is missing, unclear, or may not be the best approach, seek clarification from designers. Don't guess what accessibility information to use, or ignore including accessibility information.

Beyond working with designers to confirm how accessibility should best be implemented, there are many tools to make the process of developing accessible code quicker and more efficient and reduce the risk of introducing accessibility bugs and defects. When these tools are integrated into development environments and code management tools, building accessible code from the start and checking for accessibility issues can be optimized.

Integrating regular accessibility checks during development makes it easier to spot when an accessibility barrier has been introduced and easier to adjust code to remove the barrier. The nature of the accessibility checks may be defined by acceptance criteria or definitions of done included in product requirements, or it may be a more general set of checks that can be performed on any product. Accessibility checks take various forms, depending on the nature of the requirement. Some involve purely manual inspection, some are best performed with the aid of tools, and some can be fully automated.

We discuss accessibility testing tools and methods in more detail in Chapter 11, *Testing and Evaluation*, where we focus on the formal phase of testing and evaluating a digital product against accessibility requirements. Here we focus on methods that can be used during development for regularly checking for accessibility issues as code is written. Many of these checks map to meeting the core accessibility requirements outlined in the previous chapter, *Core Requirements*.

You may not have time to perform all of these checks whenever you write a new piece of code or edit existing code. But any effort you can make to verify accessibility will help reduce the burden on accessibility testing efforts later in the development lifecycle. These checks are intended to be applicable across multiple development platforms, except where noted.

Keyboard Interaction: Using your keyboard, you can perform several checks to identify possible issues with the keyboard user experience and, by extension, the screen reader user experience. Use the TAB key to move through each active element on the page or screen to verify:

- Every active element can be focused using the keyboard.
- Focus can move away from each active element—in other words, focus is not trapped on any element.
- There is a clear, visible indication on each element when it is focused.
- There is a logical order in which focus moves through elements on the page or screen.

Once you've performed these checks, go back and conduct further keyboard checks on each element in turn and verify that, when the element has focus, it can be activated using standard keystrokes:

- Hyperlinks on a webpage can be activated by pressing ENTER.
- Buttons can be activated by pressing SPACEBAR or ENTER.
- Select menus and comboboxes can be operated using standard keystrokes.
- Checkboxes can be toggled on and off using SPACEBAR.
- Each radio button in a group can be toggled on or off using the arrow keys.
- Text input fields can be typed into.
- Sliders can be operated using the arrow keys.
- Any element that is programmed to be disabled in certain situations is disabled when attempting to activate it using the keyboard.

Further keyboard checks can be performed on specific widget types to ensure that behavior is as standard:

- Hidden content is hidden from the keyboard. This may include content controlled by tab panel widgets, disclosure or accordion widgets, and options in menus that are hidden until they are activated by the user.
- When an action launches a dialog:
  - Keyboard focus is moved to an appropriate point in the dialog, for example, the first focusable element in the dialog or the dialog container.
  - For modal dialogs, focus is constrained within the dialog until it is dismissed.

An additional valuable keyboard check verifies the behavior of popups and tooltips is managed in an accessible way. You should supplement this check by conducting a similar check using a mouse. For any element where moving focus to and from the element causes additional content to appear and disappear, for example, in a tooltip or popup, verify that the following conditions have been met:

- The additional content can be dismissed without moving focus or hovering away from the element.
- The pointer can be moved to the additional content without the content disappearing.
- The additional content remains available until a user action dismisses it or its information is no longer valid.

Keyboard checks can be performed on webpages and applications, desktop software, and native mobile apps using a keyboard connected to a smartphone or tablet using a Bluetooth connection.

Touchscreen Interaction: Conducting accessibility checks using gestures on a touchscreen can help identify possible issues relating to touch interactions. Testing usually begins with enabling accessibility settings on the touchscreen device and then using device controls and gestures to navigate and operate the interface. For example, on a smartphone, enable VoiceOver (iOS) or Talkback (Android) and then use gestures to navigate through interactive controls. For basic navigation on iOS devices, swipe down with two fingers to read sequentially down the screen, swipe right or left anywhere on the screen to move between controls, and single-finger double-tap to activate the control that is currently focused. Verify that content and focus order are logical and sequential, and that controls are focusable and operable using gestures.

Page Content and Structure: You can conduct visual checks on whether page or screen structure is replicated in the underlying code. These checks apply to any platform using language that can convey content structure, such as HTML and native mobile applications. From a visual inspection of the content, verify that:

- Section headings are identified in code as headings.
- The hierarchy of headings is logical and reflects the content structure.
- Content that is visually structured as lists is identified as such in the markup.
- Content visually structured as a table is identified as such in markup, and that table markup reflects the table structure, including the identification of row and column headers.

Flexibility and Adaptation: For platforms that support display customization, you can check whether the way a visual design has been implemented honors display changes that a user might make. These include verifying:

- When a page is zoomed or the text size is increased, all content remains readable without disappearing or becoming obscured or overlapped by other content.
- When adjustments are made to line or letter spacing, all content inherits those adjustments and remains readable.

Accessibility Testing Tools: Tools can speed up the process of conducting accessibility code checks and improve accuracy by analyzing code for issues that may not be easy to spot visually. Tools for supporting accessibility checks come in a range of forms. For web development, browser developer tools have emerged as powerful built-in tools to support accessibility checks. Developer tools allow direct inspection of an element to understand its accessible name, role, and current state. Tools also provide a means to pause scripts during execution, allowing you to inspect an element's accessibility properties at different states and confirm whether attribute values are updated appropriately when states change.

Browser extensions and bookmarklets are another valuable source of tools for specific accessibility checks for websites and web applications. These tools can help you perform manual checks by visually highlighting specific areas of interest that might reveal an accessibility issue or perhaps showing content on a page that would otherwise be visually hidden, such as the `alt` attribute for an image or the `tabindex` value of an active element. You can find lists of bookmarklets that support accessibility checking or write your own JavaScript to perform very specific transformations of a page to support accessibility checks.

Using Assistive Technology: You may want to include limited checking of content and functionality with selected assistive technologies to complement other accessibility checks and to verify that content and functionality behave as expected. For example, checking functionality with a screen reader running can help identify any instances of content not being read out as expected or inaccurate accessibility information being communicated for a specific user interface element. This method is not intended to replace more formal assistive technology testing or evaluation with users but can provide valuable information to inform development, so long as you use the assistive technology in a realistic way.

Recording Results: Regardless of how many accessibility checks you carry out as part of development and regardless of the method of performing the check, the work is only worthwhile if you fix any issues the check identifies. Since a check is specific to one or two potential issues, isolating the code fix necessary to make the change should be straightforward.

And if it isn't, then document the issue, the predicted impact on affected users, and discuss with the product team.

10.3 Design and Development Practices

In Chapter 7, *Accessibility in Practice*, we discussed the range of accessibility roles and responsibilities and how responsibility for accessibility is a shared endeavor. While each role in the product team makes an important contribution to accessibility, much of the work on accessibility happens in design and development, where requirements are articulated in design and implemented in code. This means that design and development practices have a major role in determining whether accessibility is effectively implemented and managed.

A team that has the requisite design and development knowledge of accessible code is well placed to build accessible digital products. Equally important is the need to adapt product development team practices to optimize accessibility decision-making. Embedding accessibility into standard design and development practices can be achieved through various process adjustments. Let's look at some of these adjustments.

10.3.1 Assess Third-Party Technologies and Tools for Accessibility Support

When a product team is considering third-party technologies or tools to support development efforts, there's an opportunity to ensure the solution reduces rather than increases the accessibility burden. We can think of a tool's accessibility support in two dimensions.

First, does the tool help us create accessible code and accessible content? Choosing a development platform has implications for whether accessibility is included "out-of-the-box" in platform features and functionality or whether accessibility will need to be added to the platform as a customization. This makes it essential to prioritize accessibility support when choosing third-party tools that will be used to support building the product, so as to reduce the effort required to remediate generated code to ensure it meets accessibility requirements. Similarly, if your product will include third-party code or content, for example, to provide specific functionality like a chatbot or login verification functionality, make sure that candidate solutions are assessed for accessibility. Ideally, select the solution that has the best support for accessibility. If this isn't possible, be aware of the accessibility limitations of the chosen solution and take steps to address any barriers present.

Second, is the tool itself accessible? Product development teams include designers, developers, and content creators who have accessibility needs. Any tool the product team plans to use must be usable by all team members, and that includes people with disabilities. Also, if a tool has an interface with accessibility issues, how confident can we be that it will help us create accessible products?

Every engineer should know… disabled people are digital creators and consumers. We need accessibility in design and development tools.

By Yasmine Elglaly

People with disabilities are not solely end-users of software; they also create software and digital content. Therefore, it is crucial to consider accessibility during the development of all types of software, including software development tools like integrated development environments (IDEs), version control systems, build tools, and continuous integration/continuous deployment systems. Making software development accessible to people with disabilities can lead to a more diverse and inclusive software industry. When people with disabilities are involved in software development teams, they bring a unique perspective and firsthand experience of accessibility needs and challenges. By creating a culture of accessibility and inclusion in software development, we can empower software teams to create software that is more accessible, more representative of all people, and ultimately more valuable to society.

10.3.2 Focus Processes on Meeting Defined Accessibility Requirements

Adapting development processes to ensure that they focus on addressing formally expressed product accessibility requirements helps improve the efficiency of the process of building accessible products. Core requirements can be distilled into checklists that each team member can reference as they implement designs and functionality. When accessibility is included in acceptance criteria for a given piece of functionality, making sure these accessibility criteria have been met before code is pushed to the product codebase will help minimize the accumulation of accessibility-related technical debt and reduce the need for accessibility remediation.[5]

10.3.3 Use Design Tools That Support Manual Accessibility Checking and Annotation

Software applications for supporting visual and UX design and prototyping are increasingly providing functionality that conducts accessibility checks of designs and supports annotating designs with accessibility information. Extensions and plugins provide predefined sets of annotations that you can adapt for your own purposes, making the task of annotation more efficient and less likely to miss important information.

10.3.4 Take Advantage of Automation in Development

Development environments also provide an increasing range of tools to support accessibility checking in code creation. Wherever development processes harness automation for code quality checks and code management, there is an opportunity to include accessibility. Automation plays an important role in accessibility testing, and the process of managing code as it is developed can be significantly strengthened through automated accessibility checks. Automation can improve the accuracy of testing and reduce the time taken to perform tests. We discuss automated methods for accessibility testing further in Chapter 11, *Testing and Evaluation*.

10.3.5 Use Meetings to Plan and Review Accessibility Efforts

Collaborative approaches help solving the accessibility problems more quickly and more effectively. Where a product team follows an agile development approach, sprint planning activities are an opportunity to review stories to be developed for potential accessibility challenges. Including accessibility concerns in team discussions about upcoming work helps surface challenges and supports shared approaches to addressing challenges. Similarly, retrospectives provide an opportunity to review successes and challenges in meeting accessibility requirements in recent development work. They also provide a way to identify opportunities for teams to increase accessibility capabilities, for example, through improved tools or additional training. Any effort to foster a culture where accessibility is a shared conversation helps to create a collaborative approach and build capability to meet accessibility requirements.

10.3.6 Seek Feedback from Users with Disabilities Early and Often

Following requirements and performing manual and automated accessibility checks helps reduce the chances of introducing barriers. To complement these efforts, gathering feedback on designs and prototypes from users is an effective way to refine design approaches before starting in on development. Early evaluation can help catch shortcomings that might otherwise be difficult to rework. This also applies to accessibility, where evaluating designs with disabled people in the design phase can help validate decisions and identify issues that may not otherwise have been considered. We discuss more about methods for evaluation with people with disabilities in Chapter 11, *Testing and Evaluation*. Acting on feedback received during evaluation helps you ensure that, when designs are passed to implementation or early functional prototypes are built, the chances of accessibility barriers being baked into the product are reduced.

10.3.7 Engage Specialist Accessibility Advice

Product teams may benefit from access to specialist accessibility advice and support to help with decision-making. Although the long-term goal should be to build a baseline of accessibility skills and knowledge across each member of the team, regular access to a subject matter expert, whether within the organization or an external consultant, can help answer questions and resolve accessibility challenges that require advanced knowledge while also supporting knowledge development within the team. Including an accessibility coach as a member of the team can help identify ways to enhance team processes and build knowledge and skills within the team.

10.3.8 Document and Share Accessibility Decisions

As the team makes shared decisions on approaches to meeting accessibility requirements, it's important to capture decisions in a central knowledge base that can be consulted by the team in future similar scenarios. To maximize visibility, this information should be captured in a wiki or other shared repository of design and development decisions rather than in a separate accessibility resource. Particularly when the team is still evolving in its accessibility efforts, having a single source of accessibility decisions helps grow a shared understanding of accessibility approaches, improving consistency across the product and among the team.

10.3.9 Build and Use a Pattern Library

Documenting design and implementation accessibility decisions is important; even more powerful is building a library of user interface elements that have been specified in an accessible way. If your product team already uses a pattern library, make sure that the interface elements in the library reflect best practices in accessible design and provide appropriate accessibility guidance. Pattern libraries allow teams to reuse and, where needed, adapt patterns to meet specific functional requirements, improving efficiency in development while maintaining consistency in accessibility and the overall user experience. For a given user interface element that will be used in multiple places in the digital product, the library can define accessibility in terms of visual appearance across different states of the element, semantic and additional accessibility information about each component of the user interface element, how that element can be operated by different input devices, and guidance on implementing it on different platforms (for example, web or native mobile). A shared pattern library helps capture accessibility decisions in a way that they can be consistently followed by multiple teams, within a product development project, and across multiple related projects.

Attention to accessibility in design and development practices will result in ongoing attention to accessibility requirements throughout product development. In the next chapter, *Testing and Evaluation*, we'll cover how implementation of accessibility requirements can be validated and managed in more formal quality assurance activities.

Every engineer should know... supporting your team and engaging with communities makes a difference.

By Matthew Tylee Atkinson

When we make things, we naturally want to share them with as many other people as possible (just show a designer or developer someone struggling to use their app, and you'll have an instant convert to inclusivity—people naturally *want* to reach and empower others). It's vital to build on this to encourage and support everyone in your team to do their most accessible work.

You'll likely already know people who are passionate about enabling others to benefit from their designs and creations. They're the ones who are excited and driven. The best thing you can do as a team is simply *talk to each other*. Most importantly: talk to each other regularly, often, and take the accessibility barriers that your users have met into account when designing new iterations of your content and UI. For example, if your buttons' accessible names were missing, or inconsistent, or you weren't able to use your site via the keyboard before, make sure the designers know about the need to document those sorts of things next time around.

If you're the only accessibility champion on your team, no worries! Find others in different parts of your organization, talk to them, spread the word, and share the learning. Large accessibility programs at even larger companies have begun with one person beating the drum, and gradually being joined by others. It's great when it starts from the top, but the grass roots will always be vital to nurturing such cultural development.

Speaking of reaching out to other teams: you aren't on your own, because there are countless others of us out here who are always learning from each other— how to make the latest tech work better for people, and refining our existing approaches to accessibility. There's always more to learn and share, though you can benefit a great deal from the experiences of those who've gone before you, from accessibility consultants to fellow developers. Your local accessibility meet-up is a great place to start, and organizations such as the W3C provide a wide range of knowledge and resources that can help you. But in this industry, you also have a rare and precious opportunity to *shape* our shared resources and knowledge.

If you figured out how to make the framework you're using more accessible, share it! You can bet someone else will benefit, and then their users will too. Talk about your story (ups *and* downs) at meet-ups; it'll teach you valuable skills, and help grow our shared accessibility support network.

If you find some areas of the accessibility standards, DevTools UI, or even assistive technologies themselves could be improved, then provide that feedback; your input can and *does* make a difference, and you'll be paving the way for those who follow in your footsteps.

Takeaways

As an engineer, you should:

- Pay close attention to accessibility considerations when designing digital products and choose design approaches that have the greatest likelihood of meeting user accessibility needs. Prioritize accessibility in design and development approaches.
- Make full use of accessibility features in native platform elements and avoid introducing unnecessary custom components and code wherever possible.
- Regularly evaluate accessibility in design and development artifacts and prototypes to ensure the chosen approach to implementing requirements is working as expected.
- Work hard to build and use communication channels for accessibility decision-making across product team roles and design and development activities. Describe and document accessibility requirements, seek clarification for any unclear requirements, discuss options for approaches, and document decisions.
- Work with design and development tools that are optimized for accessibility, both through features that support accessible product development and an accessible administrative interface that can be used by team members who have accessibility needs. Avoid choosing inaccessible design and development tools that may limit contributions due to accessibility barriers.

Notes

1 Inclusive Design Research Center. *What is inclusive design?* idrc.ocadu.ca/about/philosophy
2 *Inclusive Design Principles.* inclusivedesignprinciples.org
3 S. Horton and W. Quesenbery (2014) *A Web for Everyone.* New York: Rosenfeld Media.
4 E. Tufte (1990) *Envisioning Information.* Cheshire, CT: Graphics Press.
5 C. Triplett (2023) *The Book on Accessibility: Developers.* www.thebookonaccessibility.com/roles/developers

11

TESTING AND EVALUATION

Objectives

Our objective for this chapter is to introduce effective methods for integrating accessibility into processes and practices for digital product testing and evaluation. We cover methods for testing to establish how well technical accessibility requirements have been met, and evaluation to establish the extent to which people with disabilities can successfully and independently use a digital product. We consider when in the development process accessibility testing and evaluation methods are most effective and ways to record the results of testing and evaluation that are useful to product teams.

Once you're through this chapter, you should know:

- When and how to carry out accessibility testing and evaluation.
- How to make the best use of automated testing tools to improve the efficiency and effectiveness of testing.
- Considerations when involving people with disabilities in evaluation activities.
- Key information to gather when recording testing and evaluation results.

Introduction

The purpose of software testing is to show that "a program does what it is intended to do and to discover program defects before it is put into use."[1] This can be broken down into two primary objectives for testing and evaluation:

1. To demonstrate that the digital product has met its requirements.
2. To identify defects in operation so that a plan to address those defects can be implemented.

DOI: 10.1201/9781003288060-13

Accessibility testing and evaluation therefore helps you determine whether accessibility requirements have been met for a digital product, find accessibility defects in the product, and help product teams understand what work is needed to meet requirements and address defects.

The recurring theme we hope you've picked up through this book is that product accessibility goals are best achieved through integrating accessibility-focused activities into every stage of the product lifecycle. This holds true for accessibility testing and evaluation. Formal testing before product release is an important phase of the product development process. At the same time, accessibility testing is most effective as a shared responsibility between people responsible for designing and building a digital product and people whose role is specifically to test what's built. The more often accessibility testing and evaluation is performed, the earlier in the product development lifecycle accessibility issues can be addressed. With this integrated approach, it's less likely that formal testing toward the end of the development lifecycle will identify significant accessibility barriers that are difficult or expensive to fix. In turn, there is a reduced likelihood that a product will be released with significant barriers.

This chapter addresses strategies, activities, and supporting resources to help with the process of accessibility testing and evaluation, including the critical aspect of how to report results so that product teams can use insights to improve the product. We distinguish between accessibility "testing" and accessibility "evaluation" as two related and complementary activities:

- **Accessibility testing** refers to activities that can be performed by an engineer using standard test procedures to verify whether technical accessibility requirements have been met and to identify accessibility defects.
- **Accessibility evaluation** refers to activities conducted with a focus on, and usually directly involving, users with disabilities to establish whether usability requirements have been met and whether people with disabilities can use the product for its intended purpose.

Elana Chapman, Accessibility Research Manager at Fable, describes the symbiotic relationship between accessibility and usability: "You can't have an accessible experience if it's not usable and you can't have a usable experience if it's not accessible."[2]

Generally speaking, testing is inclined toward a more objective measurement of a digital product, using formal test processes that can be automated to some extent. Evaluation generally involves more subjective assessment, though evaluation can also include the collection of objective quantitative data, such as the time taken to complete a task. Both testing and evaluation serve the evaluative purpose of accessibility testing, to assess whether the product works as intended for disabled people and to discover and record accessibility defects to address in managing accessibility.

11.1 Accessibility Testing

Accessibility standards are generally written to help make the process of testing against the standard as objective as possible, but the nature of accessibility requirements means that test procedures vary in method and output. Some test procedures involve code inspection, while others require visual inspection of an interface or screen design. Some are straightforward to perform, while others require a more in-depth comparison of the intended user experience against the actual user experience. This means that some accessibility tests may require careful guidance to ensure that they are correctly performed. It also means that some tests are more easily automated than others.

Automated accessibility testing can significantly speed up the process of testing and improve the accuracy of test results. Tools used for automated testing can also provide valuable prompts to support the process of manual testing. At the time of writing, no automated testing tool yet exists that can fully automate the process of accessibility testing. But with advances in processor power, artificial intelligence, and the ability to analyze large quantities of code for instances of issues, we expect that the proportion of accessibility tests that can be automated will likely increase in the coming years.

So, accurate and efficient testing requires a combination of automated and manual methods, some of which require a fair degree of experience and expertise to perform a test procedure and assess results. How you combine automated and manual accessibility testing is one of the most important decisions to make in your product accessibility strategy. To understand the relative merits of automated and manual testing approaches, it's helpful to consider the evolution of testing tools.

11.1.1 Testing Tools

Automated testing tools can significantly reduce the burden of performing accessibility testing by rapidly scanning multiple webpages or other digital assets, allowing humans to focus on tests that are difficult or impossible to automate. Importantly, automated tools can also support the process of manual testing by highlighting areas of a page or screen a manual tester should inspect in detail for potential issues and providing guidance on how to perform manual inspections.

Automated accessibility testing methods and tools have significantly advanced in power, quality, and flexibility since automated tools for checking webpage accessibility first emerged in the late 1990s. Early tools were web-based services that took a page URL as input and crawled the page source code, reporting on detected barriers. These early tools were able to report results in list form and also use icons to graphically identify the location of errors on the page, supporting testers in finding the code that was the source of the error. Web pages in the late 1990s tended to be static HTML and images, with little in the way of interactivity beyond links and forms. In the early 2000s, dynamic pages became more common, with interactivity provided through the browser using JavaScript. As a result, accessibility testing tools evolved away from testing page source code as delivered from the server to testing code of the interface as rendered to the user, represented in the browser's document object model (DOM).

At the time of writing a wide range of automated tools exist to support web accessibility testing. These include tools embedded in browser developer tools, standalone accessibility testing tools available as free and open source and commercial products, and tools that can be integrated as part of development frameworks and environments. In parallel with the growth in quality and quantity of tools for automating accessibility testing of web applications, automated testing tools began to appear in software development frameworks, such as Eclipse for Java application development and Microsoft's Visual Studio.

As activity increased in building mobile apps, the availability and quality of tools for accessibility testing of mobile app code also increased. Development environments for Android and iOS include a range of tools available to support accessibility testing. As with any tool to support development, the availability of automated accessibility testing tools is dynamic. New tools emerge while previously useful and popular tools are retired, neglected, or disappear.

Along with an increase in the number of tools, there has been an increase in the quality and breadth of the rules used by automated testing tools. Rulesets contain the logic used to conduct each test, and each rule in the ruleset will correspond to a specific accessibility test that can be performed by the tool. The rule will typically include a name, explanation of the test, associated requirements, for example, the WCAG guideline and success criterion, and priority level, and may also include disability groups affected. When data about a rule is stored in structured format, it can be output along with the result of a test using the rule, which for a given page or screen may be a pass, fail, or a warning that it requires a manual check.

Several accessibility testing rulesets are open source. Some accessibility testing tools offer APIs, software development kits (SDKs), and other integration capabilities that mean their rulesets and the results of testing against these rulesets can be harnessed by other tools. This gives product teams the opportunity to use rulesets in their own tools and processes, for example, by including accessibility tests in code linting processes, in continuous integration/continuous development (CI/CD) processes, and as part of formal tests run when code is committed to a master repository or pushed from one environment to another. This way, accessibility tests can be included in automated quality assurance testing and regression testing processes.

As we've noted, automated tools can't test every accessibility requirement. Relying on automated tools for accessibility testing still leaves many issues uncovered, though there is some debate over the proportion of accessibility issues that can be tested automatically versus manually. The extent to which an accessibility test can be automated depends on two main factors:

1. The level of human assessment and judgment needed to perform a test on a specific page or screen.
2. Whether that assessment and judgment can reliably be expressed in a set of heuristics that an automated tool can follow.

Consider as an example WCAG Success Criterion 1.4.1, Use of Color, where "color is not used as the only visual means of conveying information, indicating an action, prompting a response, or distinguishing a visual element." Like all WCAG success criteria, SC 1.4.1 is written as a statement that is either true or false for a given digital product. To find out if a webpage meets the criterion, a human tester can examine the page for instances where color is the only way that information is conveyed and, based on their findings, decide whether the webpage satisfies the requirement.

SC 1.4.1 is an example of a success criterion that lends itself to manual inspection. A skilled tester can use their knowledge of ways in which this criterion can be met to inspect a page for examples of content that might be non-conformant, for example, the use of a change in text color to indicate items that are sold out in a list of products. An automated test could be written to inspect HTML and CSS code to identify where specific text is presented in a different color from the surrounding text and perform some form of content analysis to identify instructional text that makes reference to color. But the intelligence and processor power required to inspect every page element for every possible instance of potential reliance on color makes fully automated testing of this SC very challenging with current testing tools.

Some WCAG SCs require a mix of automated and manual testing. For example, SC 2.4.2 Page Titled is satisfied when "web pages have titles that describe topic or purpose." An automated test can scan a webpage for the presence of a `<title>` element and validate that the element is in the document `<head>` element and is not empty. If these conditions are met, the test passes, and the tool asserts that the SC has been met for that webpage. If these conditions are not met, the tool can confidently report that the test has failed. But to truly verify that the SC has been met, a good tool needs to go a step further and provide the tester with a prompt to manually verify that the content of the `<title>` element is a succinct and accurate description of the page content. In this way, SC 2.4.2 is an example of where evaluation can be partially automated to test some conditions, but other conditions require manual verification. Advances in artificial intelligence may increase the quality and quantity of some automated accessibility test procedures, reducing the need for manual evaluation in instances like these.

Because of the varying nature of WCAG SCs, automated tools will typically rely on a hybrid approach of fully automated testing of some SCs and partial testing of others, providing prompts for manual verification of automated test findings to establish whether an issue is genuinely an issue. Tools may also check for issues that are not standards conformance failures but rather failures of "best practice" approaches to accessibility. In such cases, tools may use terms such as "warnings" to identify potential issues and provide advice on further investigation. This means that an automated testing tool can be an excellent companion to support manual testing, lending itself to a hybrid approach to accessibility testing.

When considering accessibility testing tools, look for how well the tool identifies issues and supports testers in addressing issues, including:

- Number of false positives, or instances where a tool incorrectly reports that an accessibility issue is present.
- Number of false negatives, or instances where a tool fails to identify an accessibility issue that it claims to be able to identify.
- Quality of information provided about each issue, such as the location and nature of the issue, the relevant standards and guidelines related to the issue, and advice on how to verify whether a suspected issue is genuinely an issue.
- Support for how to fix any issues identified in the evaluation.

Also, consider how the tool can be integrated into the product development lifecycle. The more flexible a tool is at integrating testing and reporting results within the stages in the development lifecycle where they are most useful, the more powerful it will be in improving the efficiency and effectiveness of accessibility testing.

Every engineer should know… you can't rely on testing tools. They can test only 20%–30%.

By Makoto Ueki

We need to clarify the accessibility requirements that the content should ensure in the requirements definition phase. Then we design and develop the content following the requirements. Once the content is developed, we need to verify that it meets the requirements set forth in the final quality assurance phase.

There are many ways to verify that the requirements are met, and testing tools are essential for streamlining the process. However, when testing the accessibility of content, there is something you should know before using a testing tool.

It is a common theory that only 20%–30% of all web accessibility cases can be checked with a testing tool.

Let's take one specific example of a case that cannot be determined by testing tools alone. Web accessibility guidelines include the criterion, "Provide alternative text that conveys equivalent information for an image." A testing tool can automatically check for the following:

- Look for an <img> element in the HTML code.
- Check if the <img> element has an alt attribute.
- Check if there is some string as the value of the alt attribute.

However, you must check the alternative text described as the alt attribute value to determine if it is appropriate as the text alternative for the image. No testing tool can reliably do this part.

Even if it is the same image, the appropriate alternative text may vary depending on the preceding and following content and context. For this reason, you should also check where the image you are checking is used on the web page.

Unfortunately, technology at this time does not allow us to automatically determine if the alternative text wording is appropriate.

Another example is the criterion, "Make information and relationships on the web page machine readable." An example is headings—headings on a web page should be properly marked up with HTML heading elements. What a testing tool can do is check for the following:

- Find the <h1> to <h6> elements in the HTML code.
- Check for skipped heading levels.

However, where on the web page the headings to be checked are located must first be visually confirmed. In other words, the checking tool cannot mechanically and automatically find the target to be checked in the first place.

These are criteria that seem relatively easy to check with a testing tool. But you can see that, even so, not everything can be automated. Most of the work to check the accessibility of a web page requires human visual inspection, keyboarding, and assistive technology to make sure it meets the requirements.

Therefore, even if a check tool gives a score of more than 90 out of 100, you'll find many issues while you are performing a complete human check.

Testing tools should be used proactively as a tool to improve your work efficiency. However, it is important to keep in mind that the testing tool can only improve the efficiency of a limited part of the entire process.

11.1.2 Testing with Assistive Technologies

One form of manual accessibility testing is compatibility testing with assistive technologies, or AT testing. In the same way that testing might establish whether a digital product meets design, functional, and performance requirements across a range of supported platforms, such as operating systems, devices, and browsers, you can test to ensure these requirements are met for supported assistive technologies, including screen readers, screen magnification, speech input, and switch-based input devices.

Testing with diverse assistive technologies helps you verify whether the product works as expected for assistive technology users. This is important because browsers and assistive technologies can vary in how they support accessibility features. You may find that what works in one combination may not work as well in another, even if you follow an accessible design pattern that meets accessibility standards. When you test a product to verify that functionality works with a range of specified assistive technologies, you can identify where a solution may not work as expected and provide information for the product team to decide how best to address the issue and for the support team to determine how best to support users who encounter issues.

Testing with assistive technologies works best when integrated into manual test procedures rather than as a separate activity. AT testing requires competence in the use of assistive technology to reduce the chances that results are adversely affected by a tester's unfamiliarity with using an assistive technology. Hiring people with disabilities who use assistive technology as testers is one way to ensure accurate results, and teams can also engage people with disabilities who are regular users of an assistive technology in evaluating the product.

11.1.3 Commissioning an Audit

A common service performed by digital accessibility consultants is standards conformance audits. Hiring an expert third party to perform the audit helps reduce bias and increase the accuracy of results, lending greater credibility when it comes time to report on a product's conformance, for example, in an Accessibility Conformance Report (ACR). In fact, some locations, markets, and contracts require third-party evaluation for quality attributes like accessibility rather than depending on product vendors to accurately report potential defects and issues.

An accessibility audit usually involves the vendor defining a representative sampling of content and functionality and then evaluating each item in the sample against technical standards, such as WCAG, and noting each conformance failure. The deliverable from the vendor will be some report on conformance for the sample, usually in the form of a two-dimensional spreadsheet with rows for each requirement, columns for each sample, and details about conformance in each table cell. Reports in Word or PDF format are also common, using a Voluntary Product Accessibility Template (VPAT) provided by the Information Technology Industry Council (see Chapter 12, *Documentation and Support*).

For many products, a conformance audit may be the only accessibility activity in the product development lifecycle. Conformance audits are helpful for documentation purposes and to provide credible details to include in an ACR. They are also helpful for developing an accessibility roadmap for remedying known defects in a digital product. But a technical audit does not provide insight into how people with disabilities experience a product or ways to improve the product to provide a better user experience.

When commissioning an accessibility audit, it's important to define the scope of the audit for the greatest impact and relevance. Since accessibility audits require manual evaluation, it's next to impossible to fully audit every piece of content and functionality and every possible use case of a digital product. The product owner is the best source of insight into the samples that comprise the most common and important use cases. For example, any software that requires an account registration and log-in should include those user journeys in an audit. They are the equivalent of a walkway and entrance for physical facilities, and if people can't get in the door, it doesn't matter how accessible things are on the inside. Another helpful way to define a test sample is to assess which components are prevalent and necessary to operate the interface. For example, on a website, the wayfinding features like search and menus are used throughout the site and are relied on by users to navigate. Ensuring those components are part of the test sample will be beneficial.

When commissioning an audit, you'll want to get the most value for your money. This means creating a sample that addresses the most important elements, while avoiding not repeating the evaluation of the same elements or components. There's no use testing the same banner, for example, on every screen in the sample. Good test samples test common elements separately and then exclude them from other samples where they appear. Also, include details to help make the auditor's progress through the sample straightforward. For example, to test the log-in functionality, provide the task flow details needed to work through the task. Include instructions, such as "Do test", and "Don't test," in the test sample.

11.1.4 Conducting an Accessibility Test

Accessibility testing can and ideally should take place for any artifact produced during the product development lifecycle. This includes visual designs, wireframes, and other low-fidelity prototypes intended to communicate how the product might appear, be structured, and behave, as well as implementations in functional code. Indeed, before any artifact has been created, you can perform accessibility testing and evaluation of an existing product as a preparatory task to identify barriers that a reworked or replacement product should aim to avoid and influence product requirements.

The format and purpose of an artifact will influence what accessibility testing you can perform and how you test. Accessibility testing of any static artifact not written in code, which is particularly likely to be generated early in the product development lifecycle, will exclude any accessibility tests that focus on code inspection. Instead, you can focus accessibility testing on the specific goals and format of the artifact. For example, testing a visual mockup of a screen intended to show color schemes in use will be limited to assessing the use of colors, color contrast, and other aspects of visual accessibility. Accessibility testing of text-based prototypes intended to convey the intended structure and flow of content on a screen can focus on testing for logical reading order and structure. Even if the set of accessibility tests that can be conducted on a specific artifact is limited, performing those tests helps you identify barriers early and make design changes more quickly and cheaply than if they were left until later.

Accessibility testing should be performed against defined requirements to establish the extent to which a resource meets those requirements. As discussed in Chapter 8, *Requirements Specification*, accessibility requirements, including definitions of done and acceptance criteria, are often expressed in terms of conformance to named standards such as WCAG. This means that testing is most commonly performed with reference to the same technical accessibility standards. Updates to accessibility standards have made it easier to use standards to perform objective testing; as the structure and format of WCAG have evolved in subsequent versions, one focus has been on improving the "testability" of each constituent component of the standard. WCAG supports testing by expressing requirements at a fine-grained level, but standards like Section 508 and EN 301 549 include functional performance criteria as requirements (see Table 8.1), both of which with a much broader scope and, therefore, more flexible.

Generally, it helps to use WCAG as the basis for accessibility testing and build functional performance criteria test procedures on top of WCAG testing results. So, for example, if you test against all WCAG SC that relate to accessibility issues for people without vision, then you can use the results of these tests to report against the relevant functional performance criterion for people without vision. Be aware that for products on platforms where it may not be possible to comprehensively apply WCAG, such as a closed-system kiosk, you may need to be more creative in defining test procedures for verifying functional performance criteria.

While you'll likely find WCAG to be the most common reference for accessibility testing, some content types have specific accessibility standards and guidelines that can be used for additional testing. For example, test processes for PDF documents can include testing against the PDF/UA standard.[3] Digital games can be tested against the requirements of the Games Accessibility Guidelines.[4]

11.2 Accessibility Evaluation

While a product team can perform accessibility testing to establish the extent to which a digital product meets technical accessibility requirements, you can't be sure that meeting those technical requirements will ensure that a product is usable by people with disabilities for its intended purpose. Accessibility testing is focused on performing discrete tests for specific requirements in specific code components in a specific context. Accessibility test procedures don't tend to take into account the context of use, purpose of use, or characteristics of the users who will be using the resource.

When you focus on people in evaluation activities, you can move toward establishing whether the ultimate test condition is met—that people with disabilities can successfully and independently use the product for its intended purpose in its intended context of use. Think of accessibility evaluation as measuring the level of usability for people with disabilities. What is usability? According to the usability standard ISO 9241-11, *Ergonomics of human-system interaction*, usability is "the extent to which a product can be used by specified users to achieve specified goals with effectiveness, efficiency, and satisfaction in a specified context of use."[5]

Accessibility evaluation is not a replacement for testing but instead a next step, helping you find out what works well for users and where changes could be applied to make the product easier to use. In fact, testing is an essential first step to address known barriers before exposing the product to people who have accessibility needs. There's little to be gained by having evaluation participants attempt to use a product only to encounter barriers that could have been identified through automated or manual testing processes. Instead, think of accessibility evaluation as an additional layer of testing that gets you closer to understanding if your accessibility efforts have been effective. The results of an accessibility evaluation may often lead to identifying subtle adjustments to visual or functional design that do not necessarily map to a technical accessibility standard but, if implemented, can bring measurable improvement to ease of use for disabled people and likely nondisabled people too.

The purpose and objectives of testing introduced in this chapter are a reminder that accessibility evaluation is about verifying that the product meets its accessibility requirements as well as identifying barriers to ease of use. When you evaluate a digital product with people with disabilities, you also have the opportunity to discover that the product works well for its intended purpose and can generate powerful evidence that changes don't need to be made. This might be helpful to reassure internal stakeholders looking for evidence that investment in accessibility has paid off. The same evidence might also be helpful for sales and marketing to use to demonstrate a product's accessibility to potential customers.

Evaluation also presents an opportunity to adjust the priority of unresolved issues identified through testing. If an issue turns out not to significantly affect users with disabilities from being able to use the product as intended, then you might consider deprioritizing it in the backlog of issues to fix. Accessibility evaluation can help you avoid wasting time and effort fixing barriers that don't significantly affect disabled product users' ability to complete tasks, allowing you to concentrate your efforts on addressing high-impact issues.

11.2.1 Usability Evaluation

Usability evaluation is an umbrella term for a group of activities intended to establish the level of usability of a product. Some evaluation activities involve usability specialists inspecting the interface and using their experience, sometimes aided by a set of principles or heuristics, to identify possible usability barriers. Other methods involve gathering feedback, either directly or indirectly, from representative users. Think about evaluating accessibility as applying methods for usability evaluation to focus on people with accessibility needs. In this way, you can include people with disabilities in usability evaluation activities that a product team might already be planning, or run activities that focus specifically on people with disabilities.

Perhaps the most common method for establishing a measure of a product's level of usability is to conduct a task-based usability evaluation with one or more people intended to represent the product's target audience. By asking someone to use the product to perform a set of representative tasks, you can gather data on their experience, identify how successful they were, where they encountered difficulty, and how the product's behavior may be altered to make it easier to use. By involving people with disabilities in task-based usability evaluation, you can learn how usable the product is for someone who has specific accessibility needs.

Note that usability evaluation with representative users is often referred to as "user testing." You'll frequently encounter use of this term in the profession, and while it's not inappropriate to talk about user testing, it's worth remembering that the purpose of the activity is to evaluate the usability of a digital product, not to test users, nor to engage users to perform technical accessibility testing discussed previously.

11.2.2 Conducting a Usability Evaluation with Disabled People

Usability evaluation is often thought of as something you do near the end of the product development lifecycle or after launch. It's true that it can be helpful to evaluate a mature product, so long as you have budgeted time and resources to act on any issues the evaluation identifies. But usability evaluation can also be conducted on functional prototypes, specific pieces of code developed in a sprint, or even low-fidelity prototypes—any artifact that would benefit from feedback from potential users, especially if that feedback helps you revise the product's design or behavior while it's technically easier and cheaper to do so.

One benefit of prototypes—the ease with which they can be created, adapted, and thrown away—also becomes a challenge when a prototype's format presents accessibility barriers for some people. This is particularly the case for low-fidelity prototypes like paper prototypes or visual design mock-ups intended to communicate visual appearance. In addition, prototyping applications that support interaction may not provide accessibility information to assistive technologies, making it more difficult to test with AT users.

Rather than avoid prototyping, focus evaluation on the goals of the prototype and devise ways to include people with disabilities who can evaluate the effectiveness of those goals. A design mockup exploring approaches to color, design, and typography for user interface components may not be accessible to screen reader users, but it can be evaluated for legibility and readability with people who have low vision, color deficit, and people with dyslexia or other conditions that affect reading. Also, consider adapting the format of prototypes to be more inclusive. For example, along with using wireframes for testing content, layouts, and navigation, create text-based prototypes that allow for testing of content, structure, and content order with screen reader users.[6]

Further along the product development timeline, when functional code is written and designs start to coalesce, opportunities increase for usability evaluation with people with disabilities. Don't wait for an alpha or beta release; usability evaluation can happen on more constrained functionality, whether of a specific user interface component or the output of a sprint. Usability evaluation with disabled people is best conducted once accessibility barriers are removed as much as possible from the product being tested. For barriers that can't be removed before the evaluation, design the evaluation process to minimize the chances of participants encountering these barriers.

In terms of the number of participants, asking even one disabled person to evaluate a digital product will always bring some form of benefit, so long as the product team is ready to make adjustments based on what they learn from the feedback. Inviting more than one participant means you gain perspectives from multiple individuals, and you can start to establish whether there are trends in observations and distinguish significant issues from ones that less commonly occur or have less impact. "How many people?" is often a question of time and budget—the greater the number of participants, the richer the data you'll gather, but at the same time, more participants will require more time to recruit and run a study and turn study data into actionable insights.

The ideal number also depends on the purpose of the study. If your focus is to evaluate a specific piece of functionality to find out how well it works for screen reader users, then recruit screen reader users. But if your goal is to get a broader understanding of usability across a range of accessibility profiles, then recruit from a wider range of disabilities. Consider starting by recruiting a smaller number of disabled participants for a usability evaluation of product functionality during development—say 2–3 people. And when you want to establish the usability of a mature product, recruit a larger group of participants with disabilities for a more formal evaluation that verifies that your accessibility efforts have paid off and participants can successfully use the product for its intended purpose.

The chances of receiving insightful participant feedback are increased when you ask participants to interact with a digital product in a meaningful way. An effective way to do this is to ask participants to attempt one or more representative tasks—that is, tasks you expect users of the product to be able to perform. If the evaluation is of a specific piece of functionality, focus on a task that uses that functionality. That way, you can better understand the impact of any issues encountered by users on task completion, and since successful task completion can be connected to business goals (for example, successful registrations or purchases), you can make the case for investment in fixing those issues.

If you're focused on understanding the usability of a particular piece of functionality, you might choose a task that has relatively little flexibility in terms of options for the participant. For example, "Purchase the products in your cart using the test credit card we provide" is a valid task that's unlikely to give the user too much freedom of approach. In other cases, you might be less focused on a specific component and more interested in learning how participants approach a more vague or complex task, such as, "You're interested in a cheap vacation to the Caribbean sometime in the spring. Find a hotel that suits your requirements."

Every engineer should know… you need to bake layers of accessibility testing into your process.

By Kate Kalcevich

Accessibility testing is critical to ensure that what you build will work for all users. This includes people with disabilities and people with temporary and situational limitations. For example, both having low vision or being outdoors on a sunny day makes color contrast more important.

Layering accessibility testing means using a variety of tools and approaches at different stages in the product lifecycle. It helps you to catch accessibility issues early—when it's easier to fix them.

Layer one: User research

Teams should include people with disabilities in user research to understand their needs. Ensure the questions you use to screen research participants ask about their accessibility needs. If you can't find participants this way, reach out to disability organizations or vendors that do accessibility research.

Layer two: Automated tools

Automated testing tools can find anywhere from 30% to 60% of accessibility issues. Here are some free tools to choose from:

- Browser extensions like WAVE or Accessibility Insights let you test individual web pages.
- Native app tools include Accessibility Scanner for Android and Accessibility Inspector for iOS.
- Tools like axe Core or Pa11y can be added to continuous integration testing.
- Plugins like Sa11y or Editoria11y for content management systems find accessibility issues during content editing.
- Crawlers like Purple Hats check all pages in a staging or production environment.

Layer three: Manual QA

You can integrate basic accessibility QA into your existing process. Keyboard access is an accessibility issue that isn't caught by most automated tools. It's also a big barrier for many assistive technology users. Simply stop using the mouse during your regular QA testing to find elements that aren't keyboard-accessible.

Other ways to do manual accessibility QA include:

- Set the browser zoom to 200% while doing QA to check for content reflow. Content shouldn't overlap or be obscured.
- Use dark mode in your OS and see if your site works well for people with light sensitivity.
- Check that ARIA announcements work as expected by using a screen reader to listen to them.

Layer four: Testing with users

Large organizations should budget for hiring assistive technology users to test their key task flows. Assistive technologies include hardware and software that offer different ways to use computers and smartphones—for instance, by listening to content instead of reading it or entering text with your voice instead of the keyboard. Nothing gives you greater certainty that your product will work for people with disabilities and the technologies they rely on than testing with users.

Layer five: Specialist review

If your organization has an accessibility team, ask them to do User Acceptance Testing pre-release. You can also hire vendors to audit for compliance with the international standard for accessibility, the Web Content Accessibility Guidelines (WCAG).

Where to start

You don't have to include every single layer of accessibility testing right away. Start with any one or two layers and then add more layers as you get better at accessibility testing.

Remember, the goal isn't to score high in a testing tool or even to meet a WCAG guideline, but rather to make your product or service available to all users.

A longer version of this sidebar appeared in *Smashing Magazine*: https://www.smashingmagazine.com/2021/04/bake-layers-accessibility-testing-process/

11.3 Recording and Reporting Issues

Accessibility testing and evaluation efforts are only effective if the issues encountered are addressed by the product team. Acting on the results of testing and evaluation requires deliberate decision-making, taking into account the risks of inaction, opportunities for addressing accessibility defects, and project constraints. To make good decisions, product teams need clear documentation that helps them understand each issue and its impact and identify a solution that will effectively resolve the issue. Here we explore how to report accessibility defects discovered through manual and automated testing, as well as insights and feedback from end-users, so they can be effectively managed and addressed.

11.3.1 Issue Triage

There are various sources for discovering and identifying accessibility issues, including accessibility testing and evaluation activities described earlier in this chapter. Standard software quality assurance activities, such as regular automated code checks, are likely to identify accessibility issues. However, product teams may not have unlimited resources to address all issues identified in accessibility testing and evaluation—time, availability, and expertise may be in short supply at any given time.

One key activity, therefore, is determining the severity of each issue. Issue severity allows teams to set priorities, directing potentially limited attention and resources to first addressing the issues that have the greatest impact. This triage and reporting process integrates well with issue management procedures for all functional and non-functional requirements. Approaching accessibility issues in this way makes managing accessibility issues a part of general processes for software verification and validation.

One important consideration when assigning a severity rating to an accessibility issue is determining whether the issue is a defect or bug, and if so, whether it's a barrier. We can consider a bug or defect as an aspect of the system that does not meet technical requirements, for example, where the test identifies an element that does not follow a specified pattern or include required code. This type of issue might be a coding error or omission or a standards violation. A barrier is a defect that may result in a user having significant difficulty or being unable to successfully complete a task.

Depending on the goals of testing, severity—and thus prioritization—may be determined by whether the issue identified is a defect or barrier. If the goal of the testing and quality assurance activities is standards conformance, the "barrier" designation may not be a source of prioritization. But if the goal is to ensure people with accessibility needs are able to successfully complete tasks using the digital product, issues that present barriers should be prioritized over bugs and defects that may not have a significant impact on users. Our focus in this book is on digital inclusion, which means our focus in this chapter and throughout the book is on factors that will have the greatest positive impact for disabled people. That said, one aspect of digital accessibility and issue management is determining priority based on your context. You may work in a location and sector that prioritizes regulatory compliance and, therefore, addressing issues that are violations of standards.

Relatedly, when considering the severity of an individual accessibility issue, it's important to assess how much the issue will impact a user's ability to complete a defined task. We can think of task completion as a sequence of steps that users must follow and complete successfully, for example, registering for an account, printing a boarding pass, or, for that matter, submitting an accessibility defect using issue management software. There may be one defined path and steps to follow, or multiple paths that reach the same destination. Accessibility defects along the way impact users' ability to independently follow the path and successfully complete the task. Some defects may be experienced by affected users as a mild annoyance, while others may result in barriers preventing affected users from independently completing tasks. Due to the frequency along the path, even minor issues can cumulatively result in barriers by requiring significant extra time and effort. This initial assessment of impact is key to issue management to ensure defect remediation activities can be appropriately directed.

The Barrier Walkthrough Method[7] introduced a five-point severity scale, shown in Table 11.1, which can be used to categorize issues based on predicted impact on users.

Table 11.1 Example of Issue Severity Ratings (Adapted from the Barrier Walkthrough Method)

Severity Rating	Description
Critical	The issue will likely prevent affected users from completing the task.
High	The issue may prevent affected users from completing the task independently, following the prescribed user journey. Users may be able to complete the task using an alternative path or by asking for help from others.
Medium	The issue will likely hinder affected users from making progress toward completing the task, requiring greater effort and with a greater chance of errors and missteps. Users may be able to complete the task independently but may have to make guesses and repeat actions.
Low	The issue is likely to be noticed by affected users and may marginally affect task completion. Users will be able to learn and remember how to avoid or overcome the issue.
None	The issue is not likely to be noticed by users or affect task completion.

Severity ratings are valuable factors to consider when assigning priority to fixing issues. Other factors to consider when assigning priority to fixing defects include:

- **Effort to Fix:** Some issues may be easy to fix, for example, by updating the code attributes and values of a common element in a code repository or by making minor color changes to address color contrast issues. It may make sense to give "low-hanging fruit" a higher priority, even for issues with lower severity ratings.
- **Product Lifecycle:** For issues found in legacy code, triage and priority setting must take into account whether the code is actively maintained or whether it's on a path of planned obsolescence. Remediating code and platforms that are likely to be either retired or refactored is not an ideal investment of resources, unless the issues appear in high-priority task flows and produce barriers for affected users in the target audience for the digital product.
- **Prevalence:** An issue that occurs in a template or common component that appears on every page or screen of a digital product could be a higher priority defect to fix than a single instance of an issue, especially if conformance is a priority. Prevalence has to be balanced against the severity of the issue and the criticality of the task that has the issue.

11.3.2 Documenting Issues

With accessibility, documentation of accessibility issues is a key aspect of providing accessible technology. When results from accessibility testing and evaluation are effectively documented, the chances increase that product teams will be able to interpret and act on the results in a decisive fashion.

Documenting issues from accessibility testing is relatively straightforward, given the availability of technical standards. If you're using automated testing tools, test results will generally follow a structured format, with details about the requirements tested and the result in the report. The evolution of automated accessibility testing tools includes structured data provided by a ruleset and test output. This means that automated test results can also be managed in a flexible way. For example, with web accessibility testing, you can choose to view test results by WCAG priority, WCAG success criteria or guidelines, or disability group, allowing filtering of results and supporting prioritization planning. You can export test results to bug tracking systems and output them as ACRs, supporting rapid reporting of a product's accessibility level during development.[8]

The nature of a usability evaluation means that the issues identified may be less straightforward to describe in a structured way. But for product teams to be able to act on the results, it's still important to provide as much guidance on what is wrong in order to help the process of identifying and implementing an effective solution.

Essential details to document for accessibility defects include:

- **Location:** Where the defect appears in the product.
- **Expected Behavior:** How the component is expected to appear and behave, based on the design specification and requirements documentation.
- **Actual Behavior:** How the component actually appears and behaves, including steps to replicate the defect. Screenshots and screen recordings can be helpful here.
- **Related Standards:** What specific standards are violated by the implementation, such as specific WCAG success criteria.
- **Groups Affected:** Which disability groups are affected by the defect and how are they affected. If the group includes assistive technology users, documenting the AT used along with the operating system and browser can help teams replicate and test the defect.
- **Severity Rating:** What level of impact the issue will likely have on affected users.
- **Effort Estimate:** How much effort will be required to fix the issue.
- **Ownership**: Who is responsible for fixing the defect. Keep in mind that many defects require multiple team members to fix. For instance, a writer may produce a label, a designer adds it to the mockup or prototype, a developer implements the change in the code, a QA tester validates it using tools, and a user researcher confirms it through testing with users.
- **Source:** How the issue was identified, for example, through automated testing, manual testing, usability evaluation, or end-user feedback.

Accurately documenting the issues found during testing and evaluation helps product teams act decisively and effectively on fixing issues and improving the product's accessibility. In the next chapter, we'll discuss how accessibility documentation and support helps other audiences, including people with accessibility needs.

Takeaways

As an engineer, you should:

- Include automated testing tools in your professional toolkit; understand their limitations and know when and how to use them to best effect.
- When engineering digital products, regularly conduct manual and automated testing on digital products to identify and fix barriers.
- Seek out insights from people who have accessibility needs and people with expertise using assistive technologies to validate accessibility efforts and identify opportunities for improvement.
- Support issue prioritization and resolution by carefully documenting accessibility issues that arise from testing and evaluation activities.

Notes

1 I. Sommerville (2016) *Software Engineering.* 10th Edition. Boston: Pearson Education Limited.
2 C. Morales and E. Chapman (2023) *What is the Accessibility Usability Scale & its importance for assistive tech users?* www.youtube.com/watch?v= oGG3pnhm1wE
3 *ISO 14289 PDF/UA.* pdfa.org/resource/iso-14289-pdfua
4 *Game Accessibility Guidelines.* gameaccessibilityguidelines.com
5 *ISO 9241-11:2018(en) Ergonomics of human-system interaction — Part 11: Usability: Definitions and concepts.* www.iso.org/obp/ui/#iso:std:iso:9241:-11: ed-2:v1:en
6 H. van Nues and L. Overkamp (2018) Priority Guides: A Content-First Alternative to Wireframes. *A List Apart.* alistapart.com/article/priority-guides-a-content-first-alternative-to-wireframes
7 G. Brajnik (2008) A comparative test of web accessibility evaluation methods. *Proceedings of the 10th International ACM SIGACCESS Conference on Computers and Accessibility (Assets '08).* Association for Computing Machinery, New York, NY, USA, pp. 113–120. https://doi. org/10.1145/1414471.1414494
8 D. Barrell (2020) *Agile Accessibility Handbook.* Herndon, VA: Amplify Publishing.

12

DOCUMENTATION AND SUPPORT

Objectives

Our objective for this chapter is to introduce key activities around documenting product accessibility and providing support for users affected by product accessibility defects. We establish the need for documenting accessibility, the audiences who need this information, and the information they need. We walk through methods for gathering and reporting accessibility information about a digital product. We explore strategies for minimizing the harm that might be caused by accessibility defects by supporting affected users and passing information back to the product team to address.

Once you're through this chapter, you should know:

- Approaches to documenting accessibility defects identified through testing and evaluation.
- Communication and support strategies to help mitigate the negative impact of accessibility defects.
- How to respond to feedback about accessibility defects and how to make the best use of the feedback received.

Introduction

No development process is perfect. It's likely accessibility issues will creep into design and development. Having a process for tracking accessibility issues and managing their impact on affected users is a critical part of including accessibility in the development lifecycle. Equally important is the ability to document current accessibility levels in a way that can be shared with relevant stakeholders, including people with disabilities who expect to be able to use the product to complete tasks and achieve goals.

In addition, effective communication and support for people who experience accessibility issues is essential. As an engineer, you may be involved in documenting defects so that information about issues can be effectively communicated to customers and users, in addition to informing plans for addressing issues. A strong collaboration between engineers and team members responsible for communication and support is mutually beneficial. On the one hand, you can help produce accurate communication to users on effective approaches to overcoming accessibility barriers. On the other hand, you can use insights gained from supporting customers and users to prioritize remediation efforts and identify areas for improvement.

12.1 Accessibility Documentation

Whether a product is mature and publicly available or still in development, it's important to provide documentation of the product's current state of accessibility. These details may be important to executive sponsors and other key stakeholders within your company who need to monitor the current state of accessibility for reporting and monitoring reasons. Current product customers may be required to provide evidence of accessibility status for products they use, such as, schools, colleges, and universities that are obligated to provide accessible educational programs to their students, and governmental organizations that are obligated to provide accessible programs to their citizens. Prospective purchasers and users of a product may demand details about product accessibility, including known accessibility defects, in order to decide whether to purchase the product. There are many scenarios that are well met with an accurate and detailed account of accessibility defects in a product, as well as plans to address known defects.

12.1.1 Reporting Formats

Like other quality attributes that have significant impacts, digital accessibility benefits from rigorous standards and requirements. This means that, when it comes time to report on accessibility status, you can take advantage of widely adopted documentation and reporting standards and formats.

12.1.1.1 Voluntary Product Accessibility Template (VPAT) The Voluntary Product Accessibility Template (VPAT)[1] is a documentation format provided by the Information Technology Industry Council (ITI) for reporting conformance with standards for software, hardware, electronic content, and support, collectively known as information and communications technology, or ICT. As its name implies, this documentation is voluntary for vendors to use to communicate the accessibility of their products to potential and current customers.

The VPAT was initially created to help vendors document accessibility in a standard format and to help customers in the US federal government evaluate whether an ICT product meets Section 508 regulations for ICT accessibility. The ICT Accessibility 508 Standards and 255 Guidelines "address access to information and communication technology (ICT) under Section 508 of the Rehabilitation Act and Section 255 of the Communications Act" and "require access to ICT developed, procured, maintained, or used by federal agencies." The VPAT is now available in a range of versions. Each version of the VPAT allows organizations to report against one or more key accessibility standards:

- VPAT 508 for the US Section 508 Federal standards.
- VPAT EU for the European Union's EN 301 549 standard, with "Accessibility requirements suitable for public procurement of ICT products and services in Europe."
- VPAT WCAG for governments and organizations that use WCAG or ISO/IEC 40500 as their accessibility standard.
- VPAT INT, which combines all three formats for vendors that operate across multiple regulatory dimensions.

With this range of reporting formats, engineers can select the appropriate version in accordance with the requirements of their location, markets, and contract requirements.

12.1.1.2 Accessibility Conformance Report (ACR)   When product teams use a VPAT to document a product's level of conformance with accessibility standards, the result is an Accessibility Conformance Report, or ACR. (In practice, you'll find ACR and VPAT are often used interchangeably). An ACR provides results of testing against standards, along with additional information about what was tested, how, when, and by whom. With an ACR, potential customers and users can assess the product's accessibility through a recognizable format, identify the likely impact of any issues on the ability of users with disabilities to use the product as expected, and judge whether to purchase or use that product in light of known defects. Whether a product is in development or launched, it's important to have an accurate ACR that can be shared with stakeholders who ask for evidence of accessibility. A quality ACR is an excellent way to distinguish your product from others by providing up-to-date and credible details that help stakeholders, consumers, and users understand and act on the information in their planning and decision-making.

Accessibility in product procurement

If you're involved in building a digital product, an ACR is an important artifact documenting your product's accessibility. But as an engineer, you are also a likely consumer of an ACR. You might be interested in the accessibility of a third-party product or technology you're considering using in your product development efforts. One of the easiest ways to determine whether a product has been designed and implemented with attention to accessibility is by asking the vendor for an Accessibility Conformance Report. Reviewing the quality of the information they provide and how they provide it can give you valuable insight into a vendor's attitude and approach to accessibility, as well as the accessibility of the product they offer.

But keep in mind that an ACR is only helpful when it's taken seriously. The objective isn't just to have a document attesting to the status of accessibility in a digital product. The purpose is to provide readers with details about the level of accessibility of a product and its limitations, and to enable them to use those details to make an informed decision about whether to acquire and use the product. Additionally, understanding where a product has shortcomings can inform planning for how to address issues and barriers that may be experienced by people with accessibility needs when using the product. All too often, customers ask for and receive an ACR or VPAT from a product vendor and don't review the details or evaluate their implications. They don't follow up with the vendor to ask about plans to address areas of nonconformance.

Take accessibility documentation seriously, both in the products you engineer and in the products you acquire and use.

12.1.2 Report Sources

Accessibility testing and evaluation activities are essential for identifying and addressing accessibility defects during development and also help when it comes time to produce documentation and attestations. When a product has been evaluated throughout design and development and accessibility defects are documented in a standard way during testing, these details can be a great starting point for generating an ACR for the product. Additionally, accurate documentation is likely to require some level of conformance auditing in order to credibly report any accessibility defects present in the product.

Documentation requirements can help drive accessibility efforts from the start of an engineering process. For example, the VPAT format requires details of the evaluation methods used to generate the VPAT. Keeping that in mind during testing, QA, and user acceptance testing phases ensures the process includes necessary accessibility checks and documentation of testing procedures. The VPAT format requires details of the technical standards used for testing. Having the documentation requirements in mind at the start of the design and development of, for example, a new point-of-sale system means you identify the relevant accessibility standards at the start of the project, design to meet the standards, and document any areas of nonconformance along the way.

12.1.2.1 Product Documentation Completing a VPAT may be a process of reviewing existing documentation and extracting relevant details. For example, if design and development teams use issue tracking software during development and are diligent about filing accessibility issues in a standard way, searching your issue tracking software for "accessibility" within unresolved issues should provide a set of issues to document in the product's Accessibility Conformance Report. If accessibility issues found through testing and evaluation have been effectively recorded, they can be an excellent source of details when producing an up-to-date ACR.

12.1.2.2 Conformance Audit If a product's testing and evaluation processes include an accessibility audit, the audit results can be the starting point for creating an ACR. Accessibility standards conformance audits are typically conducted by digital accessibility specialists, testing a defined sample of a digital product against accessibility requirements such as the Web Content Accessibility Guidelines. The audit results indicate whether the items in the test sample pass or fail requirements and include details about any failures or considerations. The audit may also include advice on how to remediate issues identified in the audit. This information provides the context needed to accurately represent conformance status for accessibility requirements and provide a detailed explanation.

Table 12.1 shows a selection of sample audit entries for a fictitious registration and log-in screen. The first column includes the language of the WCAG requirement, and the second and third columns show sample audit results, indicating whether the registration screen and the log-in screen pass or fail the requirement. Along with the conformance status, the results include an explanation of the reason for the status and any relevant notes. We will use this sample again later in the chapter to explore the process of moving from audit to ACR.

12.1.3 Documenting Accessibility

Whether you're creating your own product documentation or working with a digital product vendor, it's important to include essential details when producing accessibility documentation. If a vendor is providing an ACR as part of an engagement, make sure to review the ACR and check for the thoroughness and accuracy of these details.

12.1.3.1 Product Details For most digital products, with countless pages, screens, content, and interactions, it's not possible to review and document all elements for accessibility. Describing the extent of the review in the report helps stakeholders, customers, and users understand what is covered in the report and what isn't. If detailed information about

Table 12.1 Sample Audit Results for a Fictional Registration and Log-in Screen

WCAG SCs	Register for Account Screen	Log into Account Screen
1.3.1 Info and Relationships: Information, structure, and relationships conveyed through presentation can be programmatically determined or are available in text.	**Fail** Grouped inputs for date of birth (MM DD YYYY) are not programmatically associated	Pass
1.4.3 Contrast (Minimum): The visual presentation of text and images of text has a contrast ratio of at least 4.5:1. [exceptions]	**Fail** Light gray text used for input labels has insufficient contrast (1.9:1)	**Fail** Light gray text used for input labels has insufficient contrast (1.9:1)
2.4.3 Focus Order: If a Web page can be navigated sequentially and the navigation sequences affect meaning or operation, focusable components receive focus in an order that preserves meaning and operability.	Pass	Pass
2.4.6 Headings and Labels: Headings and labels describe topic or purpose.	Pass	Pass
3.3.1 Error Identification: If an input error is automatically detected, the item that is in error is identified and the error is described to the user in text.	**Fail** Input error in date-of-birth inputs (e.g., entering DD MM YY instead of MM DD YYYY) is automatically detected ("Submit" button remains inactive) but not identified visually or described in text	Pass
3.3.2 Labels or Instructions: Labels or instructions are provided when content requires user input.	**Fail** Placeholder text is used to provide input labels for date-of-birth inputs (MM DD YYYY)	Pass
3.3.3 Error Suggestion: If an input error is automatically detected and suggestions for correction are known, then the suggestions are provided to the user, unless it would jeopardize the security or purpose of the content.	Pass Note that input error in date-of-birth input is not identified visually or described in text (see 3.3.1)	**Fail** Remedy instructions not provided when user enters incorrect username or password (only "Log-in failed" message)

accessibility conformance is only available for certain features—for example, the messaging, scheduling, and to-do list features for a collaboration tool—disclose the limitations of the scope in the product description. Don't claim to be reporting on the entire product when that's not the case. Instead, use the product name and description to accurately describe the features and use cases that are covered in the report, for example, "This report covers the Messages, Schedule, and To-dos features of the product."

12.1.3.2 Evaluation Methods Describe the evaluation methods used to arrive at the conformance details. If the evaluation details result from issue documentation, describe the tests and evaluations conducted during design, development, and testing. If testing was performed following established accessibility evaluation methods, using tools or evaluation methodologies, provide details and links to the tools and methods. When working with a vendor, ask for details about their accessibility testing scope and evaluation methods.

12.1.3.3 Levels of Conformance For each relevant requirement or criterion, the VPAT format provides the following definitions for indicating levels of conformance:

- **Supports**: The functionality of the product has at least one method that meets the criterion without known defects or meets with equivalent facilitation.
- **Partially Supports**: Some functionality of the product does not meet the criterion.
- **Does Not Support**: The majority of product functionality does not meet the criterion.
- **Not Applicable**: The criterion is not relevant to the product.

Any criterion marked as "Partially Supports" or "Does Not Support" requires details about which content and functionality has issues and how those issues manifest.

Take color contrast, for example. Imagine you're documenting conformance with WCAG on a product that uses light gray text against a light background for form input labels. The color contrast between the text and background fails to meet the required 4.5:1 color contrast ratio of SC 1.4.3 Contrast (Minimum). The details in the Remarks and Explanation column might include "Color contrast for light gray form input labels is 1.9:1" as an exception. If forms comprise a large proportion of the product functionality, the level of conformance would be "Does Not Support." If input forms are minimal in the product and all other text elements meet the 4.5:1 color contrast ratio requirement, the level

Table 12.2 Sample ACR Excerpt Showing Conformance Details, Based on Table 12.1

Criteria	Conformance Level	Remarks and Explanations
1.3.1 Info and Relationships (Level A)	Partially Supports	Structure is presented programmatically with the following exception: • **Registration:** Grouped inputs for date of birth (MM DD YYYY) are not programmatically associated.
1.4.3 Contrast (Minimum)	Does Not Support	Light gray text used for input labels has insufficient contrast (1.9:1).
2.4.3 Focus Order	Supports	Focus order is consistent with visual order.
2.4.6 Headings and Labels	Supports	Headings and labels are descriptive.
3.3.1 Error Identification	Partially Supports	Visual and programmatic error messages are provided with the following exception: • **Registration:** Input error in date-of-birth inputs (e.g., entering DD MM YY instead of MM DD YYYY) is automatically detected ("Submit" button remains inactive) but not identified visually or described in text.
3.3.2 Labels or Instructions	Partially Supports	Visual form labels are provided with the following exception: • **Registration:** Placeholder text is used to provide input labels for date-of-birth inputs (MM DD YYYY).
3.3.3 Error Suggestion	Partially Supports	Visual and programmatic suggestions provided in error messages with the following exception: • **Log-in:** Remedy instructions not provided when user enters incorrect username or password (only "Log-in failed" message).

of conformance would be "Partially Supports." In this case, the Remarks and Explanations column can be used to identify the pages or screens that don't meet this requirement.

Table 12.2 shows an excerpt from an ACR that reports conformance details for the sample audit results presented in Table 12.1.

12.1.3.4 Report Dates and Updates   Given the release cycles of many digital products, an Accessibility Conformance Report will likely become outdated, perhaps even before it's completed. It's important to plan for maintaining the document. Otherwise that initial investment will become meaningless, as stakeholders, customers, and users cannot rely on the accuracy of the details in the document. Establishing a regular review and update cycle is essential to protecting the investment. One approach is to integrate ACR development and production into issue tracking so that an ACR can be readily updated by rerunning a report on current accessibility defects. No matter what your approach, be prepared to manage accessibility documentation as an ongoing and continuous task rather than something you do once and are done with it.

12.2 Communication and Support

Communication is a key factor in managing accessibility for a digital product. Product users need to know about accessibility features and issues so they can assess whether the product will meet their accessibility needs and allow them to perform the tasks they need to perform. They will want to learn about features that enhance universal usability, like compatibility with assistive technology and dedicated accessibility features such as auto-captions and transcription. They need to know what defects exist in the product to decide whether, based on their accessibility needs, the defects are workable or whether they will significantly limit users' ability to effectively and independently use the product. Customers need accessibility information to support purchasing decisions, for example, to compare accessibility features among similar products. They need details about accessibility defects, for example, to support employees using workplace tools and provide workarounds and support. They may need a detailed accessibility conformance report in order to purchase any digital product due to their location, sector, or organization.

Just like the general population, most disabled people are not software developers or accessibility specialists, and they are likely not familiar with standards like WCAG or advanced technical terms. So it's critical to communicate accessibility information in language that is meaningful to the people who are affected without assuming a high level of technical knowledge. This means an effective user and customer communication strategy is essential to successfully supporting accessibility in digital products.

12.2.1 Accessibility Statement

You may live in a location or work in a sector where organizations are required to publish an accessibility statement on their website or for their digital product. For example, in the United Kingdom, public sector bodies must publish an accessibility statement. The statement must include a statement of the level of compliance with accessibility standards (fully, partially, or not compliant). The statement must detail areas that are non-compliant and provide justification for not meeting standards. There must be a section that provides details for how people can get alternative access to content or services that are not accessible to them and a way for people to contact the organization for support and to report issues. The UK regulations also require public sector organizations to provide instructions on how to escalate a complaint if the user who experiences accessibility issues is not satisfied with the responsiveness of the organization.[2] This approach to accessibility statements provides a helpful starting point for any organization, regardless of location or obligation.

Accessibility statements often apply to the website on which they're found, but they can also be used to provide information about the digital products an organization provides. In general, an accessibility statement should include a statement of the website or product's level of conformance with accessibility standards, indicators of the parts of the website or digital product that are not conformant, what this means for affected users, alternative ways to get access to content or functionality, and who to contact for help with accessibility problems. Additional details might include:

- **Organizational Policy:** A growing number of organizations include digital accessibility in their policies. Digital accessibility might have its own policy or be included in an overarching nondiscrimination policy. Referencing the relevant organizational policy is helpful in stating the extent of organizational commitment.
- **Conformance Requirements:** Organizations often have a specific standard defined in policy or practice, such as a specific version of WCAG. These details can be helpful in defining how accessibility is measured.
- **Organizational Culture:** Accessibility efforts are stronger when they are aligned with organizational culture and values. For example, digital accessibility is a key aspect of organizational efforts to promote and protect diversity, equity, and inclusion (DEI). Without accessibility, disability equity and inclusion aren't achievable. Organizations that align digital accessibility with disability inclusion may include those details in their accessibility statement as evidence of a larger inclusion effort.
- **Contact Options:** The accessibility statement is a good place to provide details of communication channels available for people to request accessibility support and report issues. We discuss communication channels further later in this chapter.

12.2.2 Product Documentation

Providing documentation in a standardized way, such as using an ACR, is a helpful and thorough way to communicate accessibility status to users and customers. ACRs are required for some customers as evidence of the level of standards conformance for different products. They may be required in the procurement process. Providing easy access to ACRs is a form of accessibility in itself. Making clear what potential issues might be present in a product up-front means users and customers don't have to go searching for the details they need to move forward.

With its focus on reporting standards conformance, an ACR isn't the most user-friendly format. It's intended to support purchasing decisions rather than the users working with a digital product. For this audience, a better source of support is product help features and documentation.

Your product may have features that are intended to enhance support for accessibility, such as built-in text resize options, keyboard shortcuts for quick access to common functionality, a reading mode that hides distracting content, or auto-captioning features. These features are helpful only when users know they're available and how to use them.

When producing product documentation, provide details about built-in accessibility features along with other features. Provide video tutorials and walkthroughs to introduce accessibility features and teach people how to use them. Note potential accessibility issues with product features and provide suggestions for how to address limitations. Make accessibility a standard attribute that you cover in all your product documentation. By documenting accessibility features along with other product features, people who don't identify as having accessibility needs may discover features that are beneficial, such as the Invert Colors feature that reverses display colors to light against dark backgrounds. This feature helps people with vision impairments, such as light sensitivity, and is generally beneficial in contexts where the increased contrast of inverted colors increases legibility, making it easier to read.

Users may also benefit from general guidance on how to deal with known accessibility issues. For example, say your website includes PDFs, which can be difficult to use with screen reader software and may not adapt well to display settings such as enlarged text or inverted colors. Provide details for users who may encounter difficulties, such as linking to resources for working with PDFs. Provide people with accessibility needs with easy access to guidance for dealing with known issues that have workarounds.

As with most communications, it can be a challenge to keep product documentation up to date. Make updating accessibility documentation a standard part of the product lifecycle, and review and update the documentation along with each new release.

12.2.3 Help and Support

Some organizations provide a dedicated support channel for people who experience accessibility issues. Examples include a dedicated email address, a web form for reporting accessibility issues, or an "accessibility" topic category on a general contact form. Consider accessibility needs when establishing and supporting communication channels.

- **Provide Accessible Contact Options**: For example, if your help desk uses a chatbot to provide support, make sure the chat feature is accessible to people who use screen reader software and alternative input methods. Likewise with a contact form; make sure the form is correctly implemented for accessible inputs and error notification.
- **Provide Multiple Communication Channels**: Accessible communication often requires providing multiple ways to communicate with people with different accessibility needs and preferences. Some people may find it easier to get help through a call center. Others might prefer to get help through chat, messaging, or email. Provide multiple ways for people to ask questions and get answers.

In addition to needing help with using a product, people may ask for information about the accessibility features of a product. People with accessibility needs may need information about features to assist in deciding whether to purchase the product. People may need to know about accessibility features and defects to support procurement. As an engineer, you can help customer support staff by alerting them to defects and barriers that users may encounter when using digital products. When documenting accessibility issues, look for workarounds that are clear of accessibility barriers, such as an alternative way to perform the same function when there's a control that isn't operable using the keyboard. Report those workarounds to support staff so they can provide them to users who encounter issues.

Also, partnering with support staff can help with managing accessibility since they hear directly about accessibility issues with the product. Look for ways to integrate client and customer support with bug reporting and tracking. That way, support staff know when they are responding to a known issue and can access bug reports to assist in providing support. They can add additional details from the exchange to the record to assist developers in addressing the issue. This exchange can be facilitated by something as simple as adding "accessibility" as a keyword in the bug and help tracking systems and using that keyword to tag accessibility-related issues.

Every engineer should know... inaccessible design is the problem, not the user who raises the issue.

By Erich Manser

Another thing I'd encourage engineers who are concerned with digital accessibility to understand is that when barriers exist, the issue lies with the design, and not with the person experiencing the barrier.

This may seem obvious, but too often when someone reports an inability to access something, the first reaction is to point to the person's disability, whether blindness, deafness, or something else, as the source of the difficulty.

However, ability is a spectrum, and we all recognize that part of the human condition involves people being at various points along that spectrum. This means that digital content or other products that are designed in ways that ignore this reality are really failures of design.

For example, in recent years we've likely all noticed an increase in the use of QR codes, which are those seemingly meaningless squares of black and white block images, often contained in advertising or promotional materials across print, television and digital media.

As a person with some remaining vision, however, I have also noticed there are times when digital QR codes are presented silently on the screen, appearing briefly to prompt some action to be taken, before disappearing again. There is often no indication given of their presence, other than visual. In these cases, there is no way for a non-visual person to perceive the QR codes, nor to act upon them.

Clearly, people who decide to use digital technologies in this way realize that blind people exist. When confronted with the fact that some with disabilities are unable to participate equally in the benefits of such technologies, responsibility is often turned back on the disabled person, with such claims as "I didn't think blind people watch TV" or "they should have someone tell them when a QR code is there."

Frankly, this should not be acceptable to any of us. We must do better, and with the promise of today's technologies, there is simply no reason why we can't.

12.2.4 Responding to Feedback

Customers and users can be a source of helpful insights into accessibility defects that have significant impact, given that the defects are significant enough to report and seek support. Accessibility feedback may come from:

- People with disabilities, either directly or indirectly. Feedback may come through channels provided by an organization, such as the help desk, customer service, or through word-of-mouth. Feedback about product accessibility may also appear on social media.
- Accessibility specialists providing feedback that may not have been directly solicited.
- Potential customers doing due diligence on products they're considering purchasing or using.
- Lawyers, acting for plaintiffs alleging disability discrimination.

In the vast majority of cases, feedback received from third parties will have a genuine basis. In particular, feedback from people with disabilities is very likely to signify that someone is unable to use the product. Gather as much information as possible about the problem, where it is, and what the solution might be. And be aware that the absence of feedback on accessibility defects does not mean that there are no defects in a product. It's quite possible that a disabled user has encountered barriers and abandoned the product without reporting the issues. It's always better to proactively take steps to improve accessibility rather than wait for complaints before acting.

Managing reported accessibility defects is one half of the work. The other half is managing communication with the person who provided the feedback, so that they and others who are affected are reassured that the feedback has been received and is being acted upon. Where possible, communicate directly with the person who provided the feedback, and let other people affected know of your plans to address the reported defect, for example, through an accessibility statement on a website or product documentation. In this communication, you should:

- Thank the person for the time taken to report the feedback and apologize for the impact of the issue.
- Outline your plans for addressing the feedback. Provide an outline of how you propose to fix the issue, and when the fix might be made. Be realistic. If you have decided that the issue is unlikely to be addressed in the short term, let people know.
- Provide advice on any workarounds or alternatives that might be available to mitigate the impact of the issue. Maybe the information or functionality is available in an alternative way using the product, or the information or functionality can be accessed by phone. A useful workaround may not always be possible, but it's worth offering as a temporary measure while you take steps to fix the defect. Be careful about relying on workarounds. They may not be accessible to all affected users and require significant staffing commitment. Having an effective workaround should not become a long-term excuse for not fixing a defect.
- Provide a commitment to providing updates on progress.

Although publicly acknowledging accessibility defects might appear to be an admission of neglect or culpability, the important thing is to let people know about defects and plans to address them. Acknowledging issues is far more helpful and will likely to lead to greater trust than ignoring feedback and acting as if defects don't exist.

Takeaways

As an engineer, you should:

- Appreciate that most, if not all, digital products will have features and tasks flows that include accessibility defects.
- Be prepared to disclose and document accessibility defects and share plans to address them.
- Understand the essential role of accessibility documentation and customer and user support in engineering digital products.
- Recognize the mutually beneficial relationship between engineering and support teams; collaborate to support customers and users and enhance accessibility.

Notes

1 ITI (2022) *The Voluntary Product Accessibility Template (VPAT)*. www.itic.org/policy/accessibility/vpat
2 Gov.UK, *Understanding accessibility requirements for public sector bodies*. www.gov.uk/guidance/accessibility-requirements-for-public-sector-websites-and-apps

13

THE FUTURE OF DIGITAL ACCESSIBILITY

One of the recurring themes of this book is anticipating and accommodating change—change in user accessibility needs, in how and why people use technology, and in what technology people use. So it seems fitting to close the book by considering what engineers should know about what might change in how we approach digital accessibility.

We have seen many trends and developments in accessibility in our time in this field, the vast majority of which have improved the quality of digital interactions for people with disabilities. We're encouraged that progress will continue to be positive, while also being wary of new developments that fail to consider accessibility as a core requirement. We also recognize the incidence of misguided attempts to apply technology solutions to accessibility challenges.

13.1 More People with Accessibility Needs

Developments in medical science and technology may seem to point to a future where the incidence of disability has significantly reduced and assistive technology has advanced to the point where the negative impact of an impairment is effectively minimized or mitigated. We argue that the opposite is more likely to occur. Medical advances will continue to preserve and improve quality of life, meaning a *greater* number of people will live with a combination of mild-to-moderate sensory, physical, and cognitive impairments. Demographic trends generally point to an increase in life expectancy. Combined with an increase in technology adoption across generations, we expect to see a larger number of users who have accessibility needs. This means the demand for engineers with digital accessibility knowledge and skills is only going to increase along with the demand for accessible digital products.

DOI: 10.1201/9781003288060-15

13.2 Some Meaningful Technology Innovation

Also, we observe that great leaps forward in assistive technology are less common than one might expect. Highly publicized and lauded announcements of exciting new technologies, from artificial intelligence to exoskeletons, promise to transform the lives of disabled people. Many of these innovations fail to deliver on multiple counts. Often, they are technologies that could be described as solutions looking for problems—technical innovations that did not emerge from direct involvement and quality engagement with the people they intend to benefit. As a result, these may require significant expense and effort on the part of the intended beneficiaries to purchase, maintain, and use them. Additionally, they may not address the most pressing user accessibility needs. These "disability dongles" detract attention from addressing the everyday challenges disabled people face with equal access and participation in the digital world.[1]

13.3 Continuous Improvements in Platforms and Products

We've also seen a continuing trend in the technology sector, where the rapid development of new technology platforms, frameworks, and products leaves accessibility considerations behind. The idea that technology innovation is too fast-paced to include accessibility from the start is likely to persist for some time yet. And as long as it does, and progress is measured by the number of new features and functionality, accessibility will continue to be treated as maintenance, to be addressed as and when time permits.

Of course, we welcome any innovation that involves disabled people and that substantially and demonstrably improves equality and inclusion in the digital world. But we expect advancement in digital accessibility will be more gradual, with continual improvements to existing technology platforms and devices and a growing number of exceptional digital products that have been designed from the start with user accessibility needs in mind. The work of defining accessibility requirements in standards and regulations will continue along with the evolution of tools, technologies, and approaches focused on helping product teams more efficiently and effectively meet user accessibility needs in design and development.

13.4 Growing Demand for Accessibility-Aware Technologists

One thing we can confidently predict is that the need for accessibility-aware software engineers will continue to grow. There will be ongoing demand for people who can balance technical accessibility skills with the ability to advocate for accessibility in product requirements, processes, and tooling.

More people with first-hand experience with disability and accessibility needs will be part of product teams, bringing valuable experience and insights to bear on the processes and tools that are used to engineer digital products. And we expect to see product teams more regularly working with disabled people when establishing and validating accessibility requirements.

Accessibility is essential and achievable in the digital world. Invest in developing your accessibility knowledge, skills, and practices, and in growing your social and professional networks.

Because *you* are the future of digital accessibility.

Note

1 L. Jackson (2022) *Disability Dongle.* blog.castac.org/author/lizjackson

FURTHER READINGS

As the community of people working in digital accessibility grows and as technology evolves, so too does the number and diversity of resources available to help inspire and guide efforts. Here we provide a starting list of books related to digital accessibility and disability inclusion that we've found especially helpful in informing our perspectives and practices.

We encourage you to look beyond this list to other resources. Seek out a mix of readings, focusing both on the how-tos of accessible digital product design, development, and maintenance and on the insights and perspectives of people who experience accessibility barriers.

Online communities are a particularly valuable and generous source of instruction, guidance, perspectives, inspiration, and support. As social media networks fluctuate in popularity and health, one helpful tip is to look for posts with the hashtags #accessibility or #a11y.[1]

Design and Disability Inclusion

Disability Visibility: First-Person Stories from the Twenty-First Century edited by Alice Wong (Vintage Press, 2020). A compendium of contributions to disability culture from authors with a wide range of disabilities, each illustrating the lived experience of disability in a world of barriers and opportunities.

Demystifying Disability: What to Know, What to Say, and How to Be an Ally by Emily Ladau (Ten Speed Press, 2021). Valuable advice on the language of disability, accessibility, and inclusion, and how to address and avoid biases and stereotypes in communication and actions.

What Can a Body Do? How We Meet the Built World by Sarah Hendren (Penguin, 2020). A collection of explorations of the lived experience of disability and technologies that can be designed and engineered to support and extend how we engage with the built environment.

Mismatch: How Inclusion Shapes Design by Kat Holmes (MIT Press, 2020). Stories of actions and attitudes that lead to exclusion in design and guidance on how to shift mindsets toward inclusive design.

Design Meets Disability by Graham Pullin (MIT Press, 2011). A thoughtful exploration of how design can embrace disability, illustrated with examples of how some of the world's great designers might approach the task of designing appealing and usable assistive technology.

Inclusive Design and Accessible UX

Doing Accessible Social Research: A Practical Guide by Danielle Aidley and Kriss Fearon (Policy Press, 2021). Guidance on how to adapt research methods to be more inclusive of people with disabilities, with methods for conducting exploratory research to inform product requirements.

Inclusive Design for a Digital World: Designing with Accessibility in Mind by Reginé Gilbert (Apress, 2019). An approach to applying inclusive design thinking at every stage of the product development process.

A Web for Everyone: Designing Accessible User Experiences by Sarah Horton and Whitney Quesenbery (Rosenfeld Media, 2014). An introduction to nine core principles of accessible user experience design, using a set of accessibility-focused personas to show how careful application of each principle can meet accessibility needs.

Accessibility for Everyone by Laura Kalbag (A Book Apart, 2017). An overview of planning for, designing, and managing accessibility, with a focus on websites and applications.

Design for Real Life by Eric Meyer and Sara Wachter-Boettcher (A Book Apart, 2016). An important look at the range of stress cases that a digital product or service needs to support, showing the alignment between accessible design and sensitive UX design for diverse and often challenging scenarios that users may encounter.

Development and Implementation

The Agile Accessibility Handbook by Dylan Barrell (Amplify Publishing, 2020). A handy guide to transforming product teams to be equipped to handle accessibility requirements through process change and skills development, with a focus on agile methods.

Beyond Accessibility Compliance: Building the Next Generation of Inclusive Products by Sukriti Chadha (Apress, 2022). Guidance on effectively integrating accessibility into the design and development process, with a focus on the accessible user experience of mobile and wearable technology and other emerging technologies.

Inclusive Components by Heydon Pickering (Smashing Magazine, 2019). Common user interface patterns for web applications, with details about accessibility needs and ways to address them through design and coding.

Form Design Patterns by Adam Silver (Smashing Magazine, 2018). Best practices on designing and implementing effective online forms and how to avoid commonly encountered accessibility issues with form interaction.

Developing Inclusive Mobile Apps: Building Accessible Apps for iOS and Android by Rob Whitaker (Apress, 2020). A through review of accessibility support on Android and iOS platforms and devices, and how core principles of accessible design and development can be applied to creating accessible and inclusive mobile apps.

Note

1 A11y is a numeronym representation for accessibility, composed of the first and last letters, "a" and "y", and the number of characters in the word (11) to make "a11y". The visual resemblance to the word "ally" is not a coincidence, since people working in accessibility are allies of disability inclusion efforts.

INDEX

Note: **Bold** page numbers refer to tables.

For Product Safety Concerns and Information please contact our EU
representative GPSR@taylorandfrancis.com
Taylor & Francis Verlag GmbH, Kaufingerstraße 24, 80331 München, Germany